Advanced Methods of Pharmacokinetic and Pharmacodynamic Systems Analysis

Biomedical Simulations Resource Workshop on "Advanced Methods of Pharmacokinetic and Pharmacodynamic Systems Analysis (1990: Marina del Rey, Calif.)

Advanced Methods of Pharmacokinetic and Pharmacodynamic Systems Analysis

Edited by
David Z. D'Argenio
University of Southern California
Los Angeles, California

Plenum Press • New York and London

Library of Congress Cataloging-in-Publication Data

Biomedical Simulations Resource Workshop on Advanced Methods of Pharmacokinetic and Pharmacodynamic Systems Analysis (1990 : Marina del Rey, Calif.)
Advanced methods of pharmacokinetic and pharmacodynamic systems analysis / edited by David Z. D'Argenio.
p. cm.
"Proceedings of the 1990 Biomedical Simulations Resource Workshop on Advanced Methods of Pharmacokinetic and Pharmacodynamic Systems Analysis, held May 18-19, 1990, in Marina del Rey, California"--T.p. verso.
Includes bibliographical references and index.
ISBN 0-306-44028-8
1. Pharmacokinetics--Congresses. 2. Drugs--Physiological effect--Congresses. I. D'Argenio, David Z. II. Title.
RM301.5.B56 1990
615'.7--dc20 91-22186
CIP

Proceedings of the 1990 Biomedical Simulations Resource Workshop on Advanced Methods of Pharmacokinetic and Pharmacodynamic Systems Analysis, held May 18–19, 1990, in Marina del Rey, California

ISBN 0-306-44028-8

Printed in the United States of America

PREFACE

This volume records the proceedings of the Workshop on Advanced Methods of Pharmacokinetic and Pharmacodynamic Systems Analysis, organized by the Biomedical Simulations Resource in May 1990. The meeting brought together over 120 investigators from a number of disciplines, including clinical pharmacology, clinical pharmacy, pharmaceutical science, biomathematics, statistics and biomedical engineering with the purpose of providing a high–level forum to facilitate the exchange of ideas between basic and clinical research scientists, experimentalists and modelers working on problems in pharmacokinetics and pharmacodynamics.

It has been my experience that in many areas of biomedical research, when a meeting of this type is held, the general attitude of those experimentalists willing to attend is one of extreme skepticism: as a group they feel that mathematical modeling has little to offer them in furthering their understanding of the particular biological processes they are studying. This is certainly *not* the prevailing view when the topic is pharmacokinetics and drug response. Quite the contrary, the use of mathematical modeling and associated data analysis and computational methods has been a central feature of pharmacokinetics almost from its beginnings. In fact, the field has borrowed techniques of modeling from other disciplines including applied mathematics, statistics and engineering, in an effort to better describe and understand the processes of drug disposition and drug response. This transfer of intellectual technology, however, has and continues to be done with a keen power of discrimination on the part of basic and clinical scientists in the field. What I have found to be especially significant, moreover, is that the general area of pharmacokinetics has also provided a return benefit to those disciplines from which it has borrowed, in that the difficult problems in pharmacokinetics have provided fertile ground for the development of new approaches and significant extensions to existing techniques of mathematical modeling, data analysis, and scientific computing.

The contributors to this volume are representative of those investigators who by addressing fundamentally important basic and clinical research problems have also provided a stimulus for development of new methodologies of modeling and data analysis in pharmacokinetics and pharmacodynamics. Other contributors are indicative of the biomathematicians, engineers and statisticians who have accepted the challenge of developing these new modeling, data analysis and computational techniques.

The book itself is divided into four sections. The first involves contributions on the physiological and biochemical basis of pharmacokinetics, including chapters on the mechanisms of oral drug absorbtion, reversible metabolic processes, novel drug delivery systems for anticancer and antiviral agents, and the significance of blood sampling site in pharmacokinetic studies. The second section addresses measurement and modeling issues in pharmacodynamics as they relate to the study of corticosteroids, intravenous anesthetics and cardioactive drugs, as well as physiological approaches to pharmacodynamic modeling. Section three on pharmacometrics includes chapters on residence time distributions, estimation with model uncertainty, population analysis with intraindividual variability, and optimal design of dosage regimens. The last section addresses problems of measurement, control and drug delivery in pharmacotherapeutics, with chapters on fiber optic sensors for detecting general anesthetics, individualizing drug therapy in renal transplant and pediatric cancer patients, and drug delivery via a computer controlled infusion pump.

I wish to thank all the authors for their excellent contributions to this volume and for their generous and enthusiastic participation in the 1990 BMSR Workshop. It is with great pleasure that I acknowledge the many contributions of Mrs. Gabriele Larmon in both the efficient organization of the Workshop and in the professional preparation of this volume. Also, I would like to thank Nicolas Rouquette for his TEXnical expertise. Finally, I wish to acknowledge the support of the Biomedical Research Technology Program of the National Center for Research Resources of NIH for its support of the BMSR and its activities.

Los Angeles, California David Z. D'Argenio

CONTENTS

PHARMACOKINETICS: PHYSIOLOGICAL AND BIOCHEMICAL BASIS

PHARMACODYNAMICS: MEASUREMENTS AND MODELS

PHARMACOMETRICS: MODELING, ESTIMATION AND CONTROL

PHARMACOTHERAPEUTICS: MEASUREMENT, CONTROL AND DELIVERY

PHARMACOKINETICS:
PHYSIOLOGICAL AND BIOCHEMICAL BASIS

PREDICTING ORAL DRUG ABSORPTION IN HUMANS: A MACROSCOPIC MASS BALANCE APPROACH FOR PASSIVE AND CARRIER–MEDIATED COMPOUNDS

Doo-Man Oh, Patrick J. Sinko* and **Gordon L. Amidon**

College of Pharmacy
The University of Michigan, Ann Arbor
and
*Therapeutics Systems Research Laboratories, Inc.

INTRODUCTION

There are several models for estimating drug absorption in humans [1–8]. Both physicochemical properties of the drug and physiological/biochemical properties of the gastrointestinal tract affect the extent and/or rate of oral drug absorption. Some of these factors include: pKa, solubility and dissolution rate, aqueous diffusivity, partition coefficient, chemical and enzymatic stability, intestinal pH, transit time, gastrointestinal motility, endogenous substances such as bile salts, and exogenous substances such as nutrients (food). The systemic availability can be further reduced by first-pass hepatic metabolism. Consequently, prediction of absorption is semi-quantitative.

In this report, a recently developed theoretical approach will be discussed for drugs that are absorbed by passive and nonpassive (carrier-mediated) processes [9–13]. The chapter will focus on the prediction of fraction dose absorbed (portal system availability) rather than bioavailability (systemic availability) since factors such as first-pass hepatic metabolism will not be addressed. Drug loss in the intestine is further assumed to occur only from absorption and not from other factors such as chemical or enzymatic instability.

MACROSCOPIC MASS BALANCE APPROACH

The theoretical approach to estimating of the extent of oral absorption is based on a steady-state macroscopic mass balance on drug in the intestine [10,13]. This model does not include a stomach compartment even though a gastric emptying rate is an important consideration for some compounds. The physical model of the small intestine is taken to be a cylinder (tube) with surface area of $2\pi RL$ where R is the radius and L is the length of the tube (Fig. 1). Assuming the difference between the rate of mass flowing into and out of the tube is equal to the rate of mass absorbed, the rate of mass absorbed across the tube wall is

$$-\frac{dM}{dt} = \left(\frac{Q}{V_L}\right)(M_{in} - M_{out}) = \int\int_S J_w \, dA \qquad (1)$$

Advanced Methods of Pharmacokinetic and Pharmacodynamic Systems Analysis
Edited by D'Argenio, Plenum Press, New York, 1991

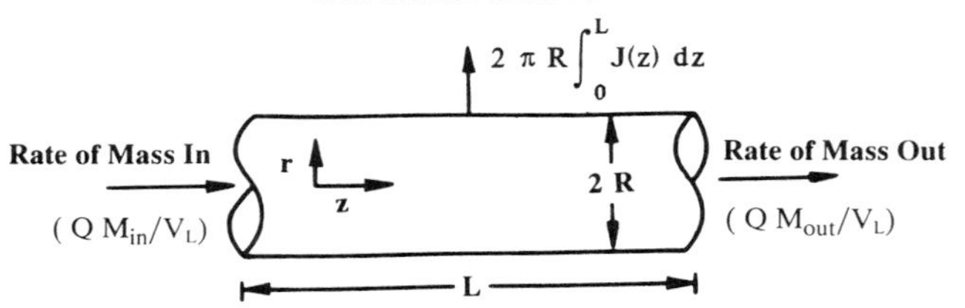

Fig. 1. Macroscopic mass balance on a tube. The rate of mass absorbed equals the difference between the rate of mass flow in and out of the tube.

where M is the mass in the tube, Q is the fluid flow rate, V_L is the luminal volume of the intestine, M_{in} and M_{out} are the mass at the inlet and outlet, respectively, J_w is the mass flux across the tube wall, and A is the absorptive surface area. The flux across the tube wall is generally expressed as:

$$J_w = P_{eff} \cdot C_L \tag{2}$$

where P_{eff} is the effective permeability [5], and C_L is the drug concentration in the tube.

With cylindrical geometry, Eq. (1) becomes:

$$-\frac{dM}{dt} = \left(\frac{Q}{V_L}\right)(M_{in} - M_{out}) = 2\pi R \int_0^L P_{eff} \cdot C_L \, dz \tag{3}$$

Further introducing the dimensionless variables,

$$z^* = \frac{z}{L} \quad \text{and} \quad C^* = \frac{C_L}{C_{in}} \tag{4}$$

where z is the axial coordinate and C_{in} is the initial concentration (M_{in}/V_L). From Eqs. (3) and (4), the fraction dose absorbed, F, is:

$$F = 1 - \frac{M_{in}}{M_{out}} = \left(\frac{2\pi RL}{Q}\right) \int_0^1 P_{eff}\, C^*\, dz^* \tag{5}$$

Eq. (5) is the basic equation for predicting drug absorption from the intestine. If the drug is passively absorbed from the intestine (constant wall permeability), Eq. (5) becomes:

$$F = \left(2\; Gz\; P^*_{eff}\right) \int_0^1 C^*\, dz^* \tag{6}$$

where $P^*_{eff} = P_{eff}(R/\mathcal{D})$, $\mathcal{D}$ is diffusivity of drug, and $Gz = \pi\mathcal{D}L/Q$ for a complete radial mixing model. The Graetz number, Gz, is the ratio of axial convection to radial diffusion times in the tube. Since both Gz and P^*_{eff} are dimensionless, Eq. (6) suggests a definition of a new dimensionless variable, the absorption number (An) [10,12,13]:

$$An = Gz \cdot P^*_{eff} \tag{7}$$

Further Eq. (6) can be rewritten as:

$$F = 2\ An \int_0^1 C^*\ dz^* \tag{8}$$

Eq. (8) suggests that there are two factors that determine the fraction dose absorbed: absorption number (An) and concentration profile in the tube.

The absorption number is a physiological parameter which is obtained by experiments. The absorption number is the ratio of radial mass transfer rate to axial convective flow rate. It indicates that An is affected by not only the permeability of a drug but also by the bulk fluid flow in the intestine. An is calculated from the intrinsic wall permeability which is estimated from single-pass perfusion experiments [10,13]. Substituting $P^*_{eff} = P^*_w\ (1 - P^*_{eff}/P^*_{aq})$ into Eq. (7) and assuming $P^*_{aq} >> P^*_{eff}$ result in:

$$An = Gz\ P^*_w \left(1 - \frac{P^*_{eff}}{P^*_{aq}}\right) = Gz \cdot P^*_w \tag{9}$$

where P^*_w and P^*_{aq} are the dimensionless wall permeability and aqueous permeability of the compound, respectively. Eq. (9) is useful because P^*_w is independent of the fluid hydrodynamics in the intestine. The Graetz number may be considered as a scaling factor, when it is applied to predicting the fraction dose absorbed in humans from experimental data in rats [13].

The evaluation of the integral in Eq. (8) requires a knowledge of how the concentration of drug in the lumen varies down the length of the tube. The concentration profile in the tube is dependent on the P^*_w, as well as the actual flow (convection) pattern and the theoretical flow model that is chosen for analysis. Two limiting cases are a mixing tank (well-stirred) model and a complete radial mixing model [10,13]. The drug concentration is also dependent upon the initial drug concentration as well as other mass transfer processes (dissolution and absorption) and/or metabolism in the intestine. Three cases need to be considered with regard to inlet and outlet drug concentrations:

$$\begin{aligned} &\text{Case I} &&: C_{in} < C_s \text{ and } C_{out} < C_s \\ &\text{Case II} &&: C_{in} \geq C_s \text{ and } C_{out} < C_s \\ &\text{Case III} &&: C_{in} > C_s \text{ and } C_{out} > C_s \end{aligned}$$

where C_{in} and C_{out} are the inlet and outlet concentrations in the intestine, respectively, and C_s is the solubility of the compound.

A complete analysis of these models has been recently developed [10,13]. Case II is the intermediate of Case I and Case III. Case I and III are described in the following sections. Figure 2 shows a theoretical plot of fraction dose absorbed versus absorption number and dose number (see below) using mixing tank and complete radial mixing models [10,13]. If a drug is transported by a carrier-mediated system, the resultant wall permeability will demonstrate concentration dependent behavior. In this case Eq. (5) should be used to predict the fraction dose absorbed instead of Eq. (8). A microscopic mass balance approach can be used to predict the fraction dose absorbed. Using simultaneous differential equations and a mass balance on a volume element of the tube, the concentration profile can be obtained to calculate the integral in Eq. (5) (Oh, unpublished results). An alternative approach to using Eq. (5) is to use a mean wall permeability [11].

CORRELATION OF F WITH An: CASE I

For the complete radial mixing model the concentration profile of a passively absorbed drug in the intestine is obtained from the mass balance at steady state:

$$C^* = e^{-2An \cdot z^*} \tag{10}$$

From Eqs. (8) and (10), the fraction dose absorbed for Case I is:

$$F_I = 1 - e^{-2An} \tag{11}$$

The fraction dose absorbed is exponentially related to absorption number. This result further supports that An is a primary parameter for predicting its absorption. The dependence of F on An is shown in Fig. 2 [11]. The Case I result is shown in Fig. 3 for the complete radial mixing model together with experimental results obtained for a number of drugs [11,12]. The correlation is excellent and can be used to estimate the extent of drug absorption for both passive and nonpassive (carrier-mediated) compounds (see below).

CORRELATION OF F WITH An AND Do: CASE III

In addition to nonpassive absorption, dose dependent absorption is also observed when a drug is dosed above its solubility. For Case III, it is assumed that

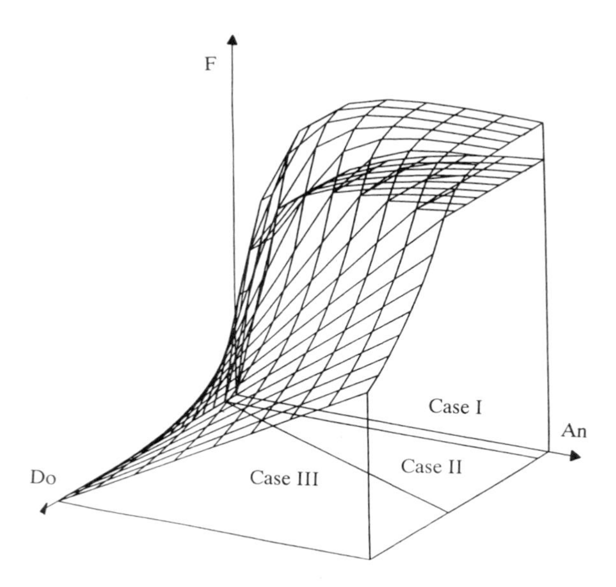

Fig. 2. Three dimensional plot of the absorption number (An) versus dose number (Do) versus the extent of absorption (F) for the complete radial mixing and mixing tank models.

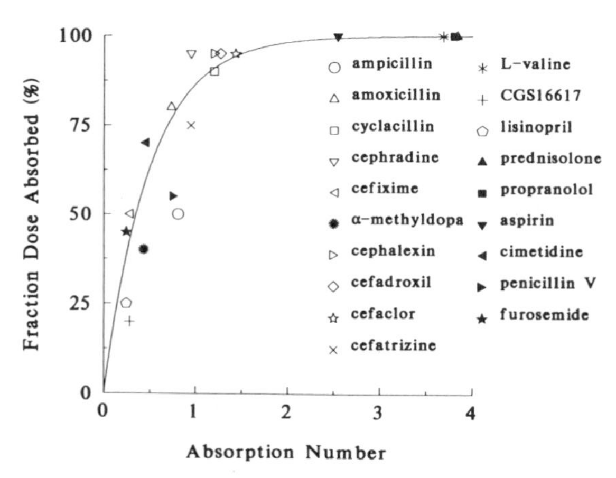

Fig. 3. Plot of fraction dose absorbed of Case I versus absorption number for several compounds whose absorption is passive or carrier-mediated. Complete radial mixing model is used for a theoretical line.

the drug concentration in the elemental volume of the tube is equal to the solubility of the drug. Substituting $C^* = C_s/C_{in}$ for the concentration profile in Eq. (8), the fraction dose absorbed for Case III is:

$$F_{III} = 2An \cdot \frac{C_s}{C_{in}} \tag{12}$$

Dose number (Do), another dimensionless parameter, is defined to be the ratio of initial concentration to the solubility of a compound:

$$Do = \frac{M_{in}/V_L}{C_s} = \frac{C_{in}}{C_s} \tag{13}$$

Therefore, the fraction dose absorbed of Case III is rewritten as:

$$F_{III} = \frac{2\ An}{Do} \tag{14}$$

From Eq. (14) the dependence of F on both An and Do is apparent. Dose number suggests that at a higher Do a lower fraction of dose is absorbed. Figure 4 shows the predicted curve for Case III illustrating the significant decrease in F as the dose number becomes larger. It suggests that dose dependency is mainly due to low solubility and the high dose taken. However, for low solubility compounds in the intestine, the luminal concentration may not be the same as their solubility due to slow dissolution and a microscopic mass balance approach for this case is being developed (Oh, unpublished results).

In both Case I and Case III, F is simply related to An, and for the intermediate case (Case II) the fraction dose absorbed can be predicted from the results of the two cases. An example of Case II is shown in Fig. 5 using the complete radial mixing model [14]. The dose dependency of amoxicillin is due to both a large dose number and a nonpassive absorption mechanism.

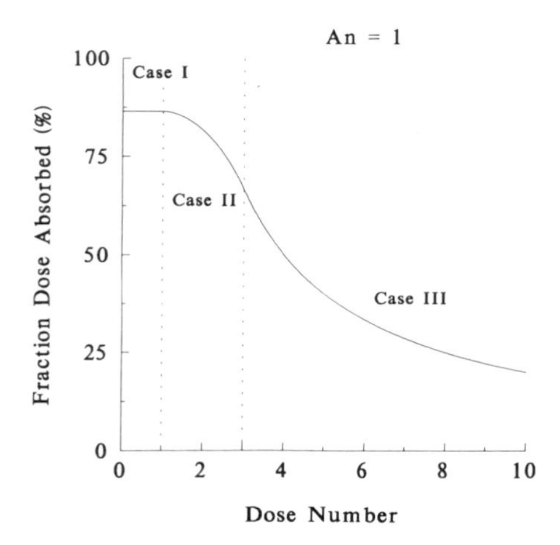

Fig. 4. Plot of fraction dose absorbed versus dose number. Complete radial mixing model is used for a theoretical line.

NONPASSIVE DRUG ABSORPTION

A modified boundary layer analysis has been developed by Johnson and Amidon [9] in order to calculate the intrinsic membrane absorption parameters from the perfused rat intestinal segment experiment. Drugs such as amino acids, dipeptides, several penicillins, cephalosporins, and ACE inhibitors are transported by a carrier-mediated (nonpassive) mechanism and have shown concentration-dependent permeabilities in rats [15–20] and absorption in humans [21]. The general expression for wall permeability is:

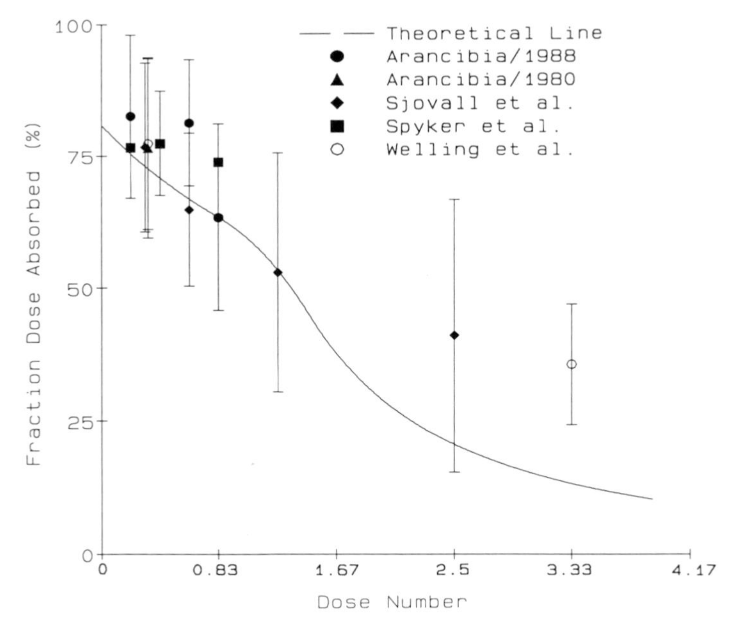

Fig. 5. Plot of the fraction dose absorbed for amoxicillin versus dose number. The curve represents the simulated curve generated using a complete radial mixing model.

$$P_w^* = \frac{P_c^*}{1 + (C_w/K_m)} + P_m^* \tag{15}$$

where P_w^*, P_c^*, and P_m^* are the dimensionless wall permeability, carrier permeability, and passive membrane permeability, respectively, K_m is a Michaelis constant for the transport system, and C_w is the wall concentration. Figure 6 shows an example of the wall permeability of ampicillin which is dependent on its concentration (Oh, unpublished results). A competitive inhibition study on the β-lactam antibiotics gives further evidence of a nonpassive absorption mechanism for these drugs [20].

To correlate F for a nonpassively absorbed drug with its wall permeability, a mean wall permeability $\overline{P_w^*}$ has been defined [11]:

$$\overline{P_w^*} = \frac{\int_{C_{in}}^{0} P_w^* \, dC_w}{\int_{C_{in}}^{0} dC_w} \tag{16}$$

where $C_{in} = M_{in}/V_L$, M_{in} is the dose administered, and V_L is the luminal volume. The absorption number is subsequently calculated from mean wall permeabilities for nonpassive compounds. Figure 3 includes compounds that are absorbed by both passive and carrier-mediated absorption pathways: nonpassively transported compounds include cephradine, cephalexin, cefatrizine, cefadroxil, cefaclor, ampicillin, amoxicillin, cyclacillin, L-valine, and α-methyl dopa.

Using a microscopic mass balance approach, the concentration profile in Eq. (6) can be obtained for estimating the fraction dose absorbed numerically. However, these results clearly indicate that the primary parameters for predicting the fraction dose absorbed are absorption number (An) and dose number (Do).

SUMMARY

Macroscopic mass balance approach has been developed and used for predicting the fraction dose absorbed in humans. The extent of drug absorption may be estimated from two dimensionless parameters: absorption number (An) and dose

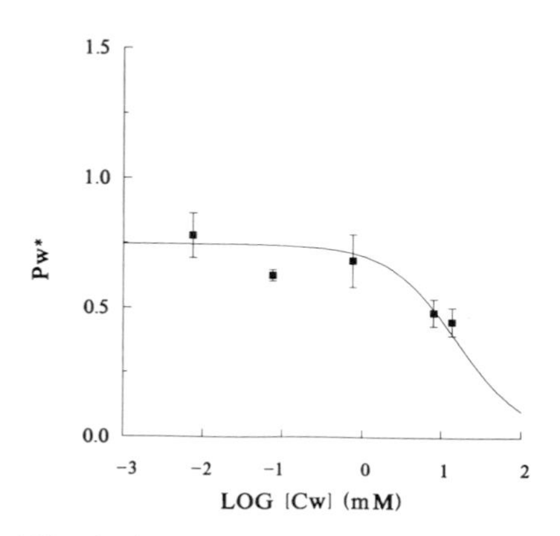

Fig. 6. Intrinsic wall permeability (P_w^*) versus wall concentrations (C_w) of ampicillin in the rat intestine: ■, experimental data (mean ± s.e.m.); ——, the best fit line.

number (Do). The absorption number is the ratio of absorption rate to convection in the intestine, whereas the dose number is the ratio of initial concentration to the solubility of a drug. The fraction dose absorbed is directly related to An and Do. Three cases with regard to inlet and outlet concentrations are discussed. This approach can be applied to compounds whose absorption mechanism is passive as well as carrier-mediated.

REFERENCES

1. N. F. H. Ho and W. I. Higuchi. Theoretical model studies of intestinal drug absorption IV: Bile acid transport at premicellar concentrations across diffusion layer-membrane barrier. *J. Pharm. Sci.* **63**:686–690 (1974).
2. G. L. Amidon, J. Kou, R. L. Elliott, and E. N. Lightfoot. Analysis of models for determining intestinal wall permeabilities. *J. Pharm. Sci.* **69**:1369–1373 (1980).
3. G. L. Amidon, G. D. Leesman, and R. L. Elliott. Improving intestinal absorption of water-insoluble compounds: A membrane metabolism strategy. *J. Pharm. Sci.* **69**:1363–1368 (1980).
4. B. C. Goodacre and R. J. Murray. A mathematical model of drug absorption. *J. Clin. Hosp. Pharm.* **6**:117–133 (1981).
5. G. L. Amidon. Determination of intestinal wall permeabilities. In W. Crouthamel and C. Sarapu (eds.), *Animal Models for Oral Drug Delivery in Man: In situ and in vivo Approaches*, American Pharmaceutical Association, Washington DC, 1983, pp.1–25.
6. J. B. Dressman, D. Fleisher, and G. L. Amidon. Physicochemical model for dose-dependent drug absorption. *J. Pharm. Sci.* **73**:1274–1279 (1984).
7. J. B. Dressman, G. L. Amidon, and D. Fleisher. Absorption potential: Estimating the fraction absorbed for orally administered compounds. *J. Pharm. Sci.* **74**:588–589 (1985).
8. J. B. Dressman and D. Fleisher. Mixing-tank model for predicting dissolution rate control of oral absorption. *J. Pharm. Sci.* **75**:109–116 (1986).
9. D. A. Johnson and G. L. Amidon. Determination of intrinsic membrane transport parameters from perfused intestine experiments: A boundary layer approach to estimating the aqueous and unbiased membrane permeabilities. *J. Theor. Biol.* **131**:93–106 (1988).
10. P. J. Sinko. Predicting oral drug absorption in man for compounds absorbed by carrier-mediated and passive absorption processes. Ph.D. dissertation, University of Michigan, Ann Arbor, 1988.
11. G. L. Amidon, P. J. Sinko, and D. Fleisher. Estimating human oral fraction dose absorbed: A correlation using rat intestinal membrane permeability for passive and carrier-mediated compounds. *Pharmaceut. Res.* **5**:651–654 (1988).
12. G. L. Amidon, P. J. Sinko, M. Hu , and G. D. Leesman. Absorption of difficult drug molecules: carrier-mediated transport of peptides and peptide analogues. In L. F. Prescott and W. S. Nimmo (eds.), *Novel Drug Delivery and Its Therapeutic Application*, John Wiley & Sons Ltd., 1989, pp. 45–56.
13. P. J. Sinko, G. D. Leesman, and G. L. Amidon. Predicting fraction dose absorbed in humans: theoretical analysis based on a macroscopic mass balance. *Pharmaceut. Res.*, in press.
14. P. J. Sinko, G. D. Leesman and G. L. Amidon. Estimating the extent of amoxicillin absorption in humans: nonpassive absorption and solubility effects. *Pharmaceut. Res.*, in press.
15. P. J. Sinko, M. Hu, and G. L. Amidon. Carrier-mediated transport of amino acids, small peptides, and their drug analogs. *J. Control. Rel.* **6**:115–121 (1987).
16. M. Hu and G. L. Amidon. Passive and carrier-mediated intestinal absorption components of captopril. *J. Pharm. Sci.* **77**:1007–1011 (1988).
17. P. J. Sinko and G. L. Amidon. Characterization of the oral absorption of β-lactam antibiotics I. Cephalosporins:determination of intrinsic membrane absorption parameters in the rat intestine in situ. *Pharmaceut. Res.* **5**:645–650 (1988).
18. D. I. Friedman and G. L. Amidon. Passive and carrier-mediated intestinal absorption components of two angiotensin converting enzyme (ACE) inhibitor prodrugs in rats: Enalapril and fosinopril. *Pharmaceut. Res.* **6**:1043–1047 (1989).

19. D. I. Friedman and G. L. Amidon. Intestinal absorption mechanism of dipeptide angiotensin converting enzyme inhibitors of the lysyl-proline type: Lisinopril and SQ 29,852. *J. Pharm. Sci.* **78**:995–998 (1989).
20. P. J. Sinko and G. L. Amidon. Characterization of the oral absorption of β-lactam antibiotics II: Competitive absorption and peptide carrier specificity. *J. Pharm. Sci.* **78**:723–727 (1989).
21. J. Sjövall, G. Alván, and D. Westerlund. Oral cyclacillin interacts with the absorption of oral ampicillin, amoxicillin, and bacampicillin. *Eur. J. Clin. Pharmacol.* **29**:495–502 (1985).

PHARMACOKINETICS OF LINEAR REVERSIBLE METABOLIC SYSTEMS

Haiyung Cheng

Department of Pharmaceutics
State University of New York at Buffalo
and
Merck Sharp & Dohme Research Laboratories

William J. Jusko

Department of Pharmaceutics
State University of New York at Buffalo

INTRODUCTION

The role of reversible metabolism in pharmacology and pharmacokinetics has been gradually appreciated. Many compounds undergo this process. For example, commonly-used drugs such as captopril [1,2], sulindac [3,4], methylprednisolone [5,6], lovastatin [7,8], procainamide [9–11], imipramine [12,13], and clofibric acid [14–16] have interconversion metabolites. Additional examples of drugs [17–31] which undergo reversible metabolism are listed in Table I. These compounds can be generally categorized according to their metabolically affected groups as: sulfides, sulfoxides, alcohols, lactones, arylamines, tertiary amines, and carboxylic acids.

Reversibility is often overlooked when the metabolite cannot be directly administered to indicate that formation of the parent drug occurs. With appreciable reversibility, the traditional pharmacokinetic methods may not reveal the true properties of the drug. Interest in the area of reversible metabolism has led to development of improved methods for pharmacokinetic analysis of drugs undergoing reversible metabolism. The purpose of this report is to review briefly the methods for obtaining meaningful pharmacokinetic parameters for this class of compounds.

METHODS FOR INTERCONVERSION ANALYSIS

Fundamental Interconversion, Elimination and Distribution Clearances

Unlike the traditional mammillary models in pharmacokinetics, the elimination of interconverting compounds is determined by two interconversion and two elimination clearances (Fig. 1). These clearance processes have been the main focus in the early attempts to obtain more meaningful pharmacokinetic parameters [32–35]. Based on mass balance relationships, the following equations for estimating

Table I: Examples of Compounds Undergoing Reversible Metabolism

Compound Class	Examples	References
Sulfides	captopril	1,2
	cimetidine	17,18
Sulfoxides	sulindac	3,4
	sulfinpyrazone	19–21
Alcohols	methylprednisolone	5,6
	cortisol	22,23
Lactones	lovastatin	7,8
	canrenone	24,25
Arylamines	procainamide	9–11
	sulfonamides	26–28
Tertiary Amines	imipramine	12,13
	chorpromazine	29,30
Carboxylic Acids	clofibric acid	14–16
	salicyclic acid	31

these clearances have been derived previously for drugs undergoing linear reversible metabolism [4,35,36]:

$$CL_{10} = \frac{Dose^p\, AUC^m_m - Dose^m\, AUC^p_m}{AUC^p_p\, AUC^m_m - AUC^p_m\, AUC^m_p} \quad (1)$$

$$CL_{20} = \frac{Dose^m\, AUC^p_p - Dose^p\, AUC^m_p}{AUC^p_p\, AUC^m_m - AUC^p_m\, AUC^m_p} \quad (2)$$

$$CL_{12} = \frac{Dose^m\, AUC^p_m}{AUC^p_p\, AUC^m_m - AUC^p_m\, AUC^m_p} \quad (3)$$

$$CL_{21} = \frac{Dose^p\, AUC^m_p}{AUC^p_p\, AUC^m_m - AUC^p_m\, AUC^m_p} \quad (4)$$

where the superscripts refer to the dosed compound, the subscripts denote the measured compound, CL_{10} and CL_{20} are the sum of all irreversible elimination clearance processes operating on parent drug and metabolite, CL_{12} is the conversion clearance of parent drug to metabolite, CL_{21} is the conversion clearance of metabolite to parent drug, AUC is the area under the plasma concentration-time curve after intravenous administration of a bolus dose of the drug or metabolite ($Dose$). Note that these equations require dual experiments where both the parent drug (p) and metabolite (m) are administered separately and plasma concentrations and AUC values of both compounds are measured.

Recently, the following equations for calculating the distribution clearances of drug (CL_{D_p}) and of metabolite (CL_{D_m}) following their iv bolus doses have also been reported [37]:

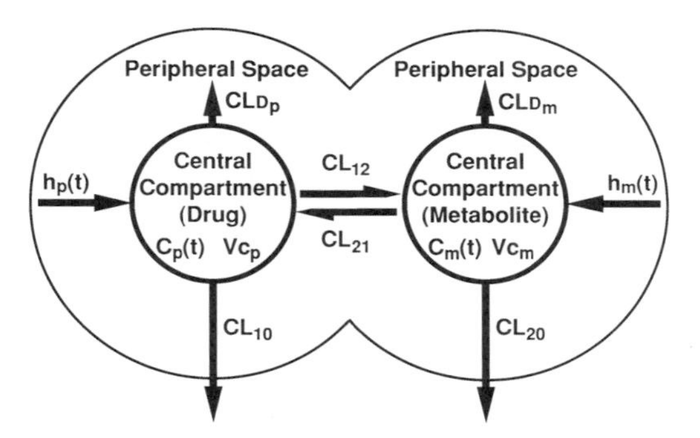

Fig. 1. Basic reversible metabolic system. $C_p(t)$ and $C_m(t)$ are plasma concentrations of parent drug and of metabolite; CL_{12} is the conversion clearance of parent drug to metabolite; CL_{21} is the conversion clearance of metabolite to parent drug; CL_{D_p} and CL_{D_m} are the distribution clearances of parent drug and of metabolite; CL_{10} and CL_{20} are the total exit clearances of parent drug and of metabolite; and $h_p(t)$ and $h_m(t)$ are the distribution functions of parent drug and of metabolite.

$$CL_{D_p} = -\left[Dose^p\, C_p^{p\prime}(0)/C_p^p(0)^2\right] - CL_{10} - CL_{12} \tag{5}$$

$$CL_{D_m} = -\left[Dose^m\, C_m^{m\prime}(0)/C_m^m(0)^2\right] - CL_{20} - CL_{21} \tag{6}$$

where $C_p^p(0)$ and $C_m^m(0)$ are the respective plasma concentrations at time zero, and $C_p^{p\prime}(0)$ and $C_m^{m\prime}(0)$ are their first derivatives at time zero. Together with Eqs. (1)–(4), these equations allow all the clearance parameters of drug and its interconversion metabolite to be calculated.

Bioavailability

In 1981, Hwang et al. [38] presented a useful theoretical basis for the estimation of bioavailability of drugs undergoing reversible metabolism. Based on this theoretical approach [38,39] and the assumption that all clearance terms remain constant, the following equation for estimating the bioavailability for oral drugs undergoing reversible metabolism (F_p) can be obtained:

$$F_p = \frac{(CL_{10}\, AUC_p^{p,po} + CL_{20}\, AUC_m^{p,po})\, Dose^p}{(CL_{10}\, AUC_p^{p} + CL_{20}\, AUC_m^{p})\, Dose^{p,po}} \tag{7}$$

where the superscript po denotes the oral administration. Similarly, the bioavailability of the interconversion metabolite after oral administration of the metabolite (F_m) can be calculated as:

$$F_m = \frac{(CL_{10}\, AUC_p^{m,po} + CL_{20}\, AUC_m^{m,po})\, Dose^m}{(CL_{10}\, AUC_p^{m} + CL_{20}\, AUC_m^{m})\, Dose^{m,po}} \tag{8}$$

Thus, as pointed out previously [38], the bioavailability assessment of a drug undergoing reversible metabolism requires plasma data from iv and oral dosing of the drug and iv administration of the metabolite in order to evaluate Eq. (7).

Volumes of Distribution

The following equations for the central (Vc) and steady-state volumes of distribution (V_{ss}) have been derived for a drug and its interconversion metabolite after separate intravenous administration of a bolus dose of both compounds [6]:

$$Vc_p = Dose^p / C_p^p(0) \tag{9}$$

$$Vc_m = Dose^m / C_m^m(0) \tag{10}$$

$$V_{ss}^p = \frac{Dose^p \left[(AUC\,_m^m)^2 \, AUMC\,_p^p - AUC\,_m^p \, AUC\,_p^m \, AUMC\,_m^m\right]}{(AUC\,_p^p \, AUC\,_m^m)^2 - (AUC\,_m^p \, AUC\,_p^m)^2} \tag{11}$$

$$V_{ss}^m = \frac{Dose^m \left[(AUC\,_p^p)^2 \, AUMC\,_m^m - AUC\,_p^m \, AUC\,_m^p \, AUMC\,_p^p\right]}{(AUC\,_p^p \, AUC\,_m^m)^2 - (AUC\,_m^p \, AUC\,_p^m)^2} \tag{12}$$

where $AUMC$ is the area under the first moment curve. The V_{ss} values can also be obtained following separate intravenous infusion at a constant rate of drug and metabolite [40]. Such equations are valid regardless of the number of linear equilibrating compartments into which the drug and metabolite distribute.

Mean Time Parameters

In addition to volume and clearance terms, the disposition of drugs can also be described by the use of the mean time parameters such as the mean residence times of drugs in the body (MRT) and in the central compartment ($MRTc$). For drugs undergoing reversible metabolism, the following equations for the MRT and $MRTc$ for drugs and their interconversion metabolites have been derived [41,42]:

$$MRT\,_p^p = V_{ss}^p \, AUC\,_p^p / Dose^p \tag{13}$$

$$MRT\,_m^p = V_{ss}^m \, AUC\,_m^p / Dose^p \tag{14}$$

$$MRT\,_p^m = V_{ss}^p \, AUC\,_p^m / Dose^m \tag{15}$$

$$MRT\,_m^m = V_{ss}^m \, AUC\,_m^m / Dose^m \tag{16}$$

$$MRTc_p^p = Vc_p \, AUC\,_p^p / Dose^p \tag{17}$$

$$MRTc_m^p = Vc_m \, AUC\,_m^p / Dose^p \tag{18}$$

$$MRTc_p^m = Vc_p \, AUC\,_p^m / Dose^m \tag{19}$$

$$MRTc_m^m = Vc_m \, AUC\,_m^m / Dose^m \tag{20}$$

Though useful for describing time constants associated with reversible biotransformation of drug, the MRT and $MRTc$ are not functions of interconversion kinetics alone but are also influenced by the distribution and elimination processes. This fact implies the need for two additional parameters which can be used separately to describe the pharmacokinetics of tissue distribution and reversible biotransformation. As pointed out previously [37], the mean transit times of drugs and metabolites in the peripheral tissues (MTT_{Tp} and MTT_{Tm}) and mean interconversion times of drugs and metabolites (MIT_p and MIT_m) may satisfy this need. These four parameters can be calculated as follows [37]:

$$MTT_{Tp} = \frac{\dfrac{(AUC_m^m)^2\, AUMC_p^p - (AUC_m^p\, AUC_p^m\, AUMC_m^m)}{(AUC_p^p\, AUC_m^m)^2 - (AUC_m^p\, AUC_p^m)^2} - \dfrac{1}{C_p^p(0)}}{-\dfrac{C_p^{p\prime}(0)}{C_p^p(0)^2} - \dfrac{AUC_m^m}{(AUC_p^p\, AUC_m^m) - (AUC_m^p\, AUC_p^m)}} \tag{21}$$

$$MTT_{Tm} = \frac{\dfrac{(AUC_p^p)^2\, AUMC_m^m - (AUC_p^m\, AUC_m^p\, AUMC_p^p)}{(AUC_p^p\, AUC_m^m)^2 - (AUC_m^p\, AUC_p^m)^2} - \dfrac{1}{C_m^m(0)}}{-\dfrac{C_m^{m\prime}(0)}{C_m^m(0)^2} - \dfrac{AUC_p^p}{(AUC_p^p\, AUC_m^m) - (AUC_m^p\, AUC_p^m)}} \tag{22}$$

$$MIT_p = \frac{AUC_p^p\, AUC_m^m - AUC_m^p\, AUC_p^m}{AUC_p^m\, C_m^m(0)} \tag{23}$$

$$MIT_m = \frac{AUC_m^m\, AUC_p^p - AUC_p^m\, AUC_m^p}{AUC_m^p\, C_p^p(0)} \tag{24}$$

As pointed out previously [37] and according to Eqs. (21)–(24), the MTT_T parameters are independent of the elimination process and intrinsic to the occurrence of tissue distribution, while the MIT parameters are independent of the elimination properties and intrinsic to the reversible biotransformation process. The MIT parameters are also indicators of persistency of drug (or interconversion metabolite) presenting as its metabolic partner in the central compartment or in the body. Moreover, they allow the time dependency of and effects of disease states on metabolic interconversion to be measured.

Finally, the following equations for calculating the mean transit times of drugs and metabolites ($MTTc_p$ and $MTTc_m$) in the central compartment have also been derived [37]:

$$MTTc_p = Vc_p/(CL_{10} + CL_{12} + CL_{Dp}) \tag{25}$$

$$MTTc_m = Vc_m/(CL_{20} + CL_{21} + CL_{Dm}) \tag{26}$$

These time parameters define the average interval of time spent by a molecule from its entry into the compartment to its next exit.

Recycling Numbers and Degree of Exposure Enhancement

Three paired recycling numbers can be calculated for drugs undergoing reversible metabolism [37]. These are the numbers of recycling of drug (R_{Tp}) and metabolite (R_{Tm}) through the tissue compartment, the numbers of recycling of drug (R_{Ip}) and metabolite (R_{Im}) through the reversible metabolism process, and the total numbers of recycling of drug (R_p) and metabolite (R_m) through the central compartment. These parameters, which can be calculated from clearances, allow mean residence times to be closely related to mean transit and interconversion times. In addition, the degree of exposure enhancement of a drug (EE_{Ip}) and its metabolite (EE_{Im}) afforded by reversible metabolism [6] can also be calculated as follows:

$$EE_{Ip} = EE_{Im} \tag{27.a}$$

$$= 1 + \frac{CL_{12}\,CL_{21}}{CL_{11}\,CL_{22} - CL_{12}\,CL_{21}} \tag{27.b}$$

$$= 1 + R_{Ip} \tag{27.c}$$

$$= 1 + R_{Im} \tag{27.d}$$

DISCUSSION

It should be noted that the derivations of all the above equations were based on linear reversible metabolic systems and, thus, are limited to this case. The processes of elimination and metabolic interconversion of the drug and its interconversion metabolite were also assumed to occur in the central compartment. The mean time approach described above has been applied to the analysis of the interconversion between methylprednisolone and methylprednisone in rabbits [37].

Recently, equations for *MRT* and *MRTc* have been derived for drugs undergoing reversible metabolism and eliminated nonlinearly [43]. These equations are identical to Eqs. (13)–(20), making them meaningful regardless of the linearity of elimination of the drug and its interconversion partner.

Efforts in the area of reversible metabolism have generated methods for calculating the fundamental clearances [32–36], bioavailability [38,39], central and steady-state volumes of distribution [6,40], mean residence times, mean transit times, mean interconversion times, recycling numbers [37], and exposure enhancement [6]. These parameters can be obtained from the doses, and the calculated values of the $C(0)$, $C'(0)$, AUC, and $AUMC$ generated from plasma concentration-time curves of drug and metabolite administered separately. These methods are easy to use, allow comprehensive analysis of linear reversible metabolic systems, but require a purposeful experimental approach to obtain appropriate data. The presence of the types of functional groups indicated in Table I should alert the kineticist to the need and opportunity to apply the described methods.

ACKNOWLEDGMENT

This work was supported in part by Grant 41037 from the National Institutes of General Medical Sciences, NIH.

REFERENCES

1. K. J. Kripalani, A. V. Dean, and B. H. Migdalof. Metabolism of captopril-L-cysteine, a captopril metabolite, in rats and dogs. *Xenobiotica* **12**:701–705 (1983).

2. O. H. Drummer and B. Jarrott. Captopril disulfide conjugates may act as prodrugs: Disposition of the disulfide dimer of captopril in the rat. *Biochem. Pharmacol.* **33**:3567–3571 (1984).

3. K. C. Kwan and D. E. Duggan. Pharmacokinetics of sulindac. *Acta. Rhumatologica Belgica* **1**:168–178 (1977).

4. D. E. Duggan, K. F. Hooke, and S. S. Hwang. Kinetics and disposition of sulindac and metabolites. *Drug Metab. Dispos.* **8**:241–246 (1981).

5. W. F. Ebling, S. J. Szefler, and W. J. Jusko. Methylprednisolone disposition in rabbits. Analysis, prodrug conversion, reversible metabolism, and comparison with man. *Drug Metab. Dispos.* **13**:296–304 (1985).

6. W. F. Ebling and W. J. Jusko. The determination of essential clearance, volume, and residence time parameters of recirculating metabolic systems: The reversible metabolism of methylprednisolone and methylprednisone in rabbits. *J. Pharmacokin. Biopharm.* **14**:557–599 (1986).

7. R. J. Stubbs, M. Schwartz, and W. F. Bayne. Determination of mevinolin and mevinolinic acid in plasma and bile by reversed-phase high-performance liquid chromatography. *J. Chromatogr.* **383**:438–443 (1986).

8. D. E. Duggan, I.-W. Chen, W. F. Bayne, R. A. Halpin, C. A. Duncan, M. S. Schwartz, R. J. Stubbs, and S. Vickers. The physiological disposition of lovastatin. *Drug Metab. Dispos.* **17**:166–173 (1989).

9. J. M. Strong, J. S. Dutcher, W.-K. Lee, and A. J. Atkinson, Jr. Pharmacokinetics in man of the N-acetylated metabolits of procainamide. *J. Pharmacokin. Biopharm.* **3**:223–235 (1975).

10. T. L. Ding, E. T. Lin, and L. Z. Benet. The reversible biotransformation of N-acetylpro-cainamide in the rhesus monkey. *Arneizmittel Forsch.* **28**:281–283 (1978).

11. K. S. Pang, J. C. Huang, C. Finkle, P. Kong, W. F. Cherry, and S. Fayz. Kinetics of procainamide N-acetylation in the rat in vivo and in the perfused rat liver preparation. *Drug Metab. Dispos.* **12**:314–321 (1984).

12. M. H. Bickel. The pharmacology and biochemistry of N-oxides. *Pharmacol. Rev.* **21**:325–355 (1969).

13. A. Nagy and T. Hansen. The kinetics of imipramine N-oxide in man. *Acta. Pharmacol. Tox.* **42**:58–67 (1978).

14. E. M. Faed. Properties of acyl glucuronides: Implications for studies of the pharmacokinetics and metabolism of acidic drugs. *Drug Metab. Rev.* **15**:1213–1249 (1984).

15. P. J. Meffin, D. M. Zilm, and J. R. Veenendaal. Reduced clofibric acid clearance in renal dysfunction is due to a futile cycle. *J. Pharmacol. Exp. Ther.* **227**:732–738 (1983).

16. B. C. Sallustio, Y. J. Purdie, D. J. Birkett, and P. J. Meffin. Effect of renal dysfunction on the individual components of the acyl-glucuronide futile cycle. *J. Pharmacol. Exp. Ther.* **251**:288–294 (1989).

17. S. C. Mitchell, J. R. Idle, and R. L. Smith. Reductive metabolism of cimetidine sulphoxide in man. *Drug Metab. Dispos.* **10**:289–290 (1982).

18. S. C. Mitchell and R. H. Waring. The fate of cimetidine sulphoxide in the guinea pig. *Xenobiotica* **19**:179–188 (1989).

19. H. A. Strong, A. G. Renwick, and C. F. George. The site of reduction of sulphinpyrazone in the rats. *Xenobiotica* **14**:815–826 (1984).

20. B.-S. Kuo and W. A. Ritschel. Pharmacokinetics and reversible biotransformation of sulfinpyrazone and its metabolites in rabbits. I. Single-Dose study. *Pharmaceut. Res.* **3**:173–177 (1986).

21. B.-S. Kuo and W. A. Ritschel. Pharmacokinetics and reversible biotransformation of sulfinpyrazone and its metabolites in rabbits. II. Multiple-Dose study. *Pharmaceut. Res.* **3**:178–183 (1986).

22. A. Kowarski, B. Lawrence, W. Hung, and C. J. Migeon. Interconversion of cortisol and cortisone in man and its effect on the measurement of cortisol secretion rate. *J. Clin. Endocr.* **29**:377–381 (1969).

23. I. E. Bush and B. B. Mahesh. Metabolism of 11-oxygenated steroids. *Biochem. J.* **71**:718–742 (1959).

24. A. Karim, J. Zagarella, J. Hribar, and M. Dooley. Spironolactone. I. Disposition and metabolism. *Clin. Pharmacol. Ther.* **19**:158–169 (1976).

25. S. Asada, T. Ohtawa, and H. Nakae. Reversible pharmacokinetic profiles of canrenoic acid and its biotransformed product, canrenone in the rat. *Chem. Pharm. Bull.* **38**:1012–1018 (1990).
26. M. Shimoda, E. Kokue, T. Shimizu, R. Muraoka, and T. Hayama. Role of deacetylation in the nonlinear pharmacokinetics of sulfamonomethoxine in pigs. *J. Pharmacobio-Dynam.* **11**:576–582 (1988).
27. J. G. Eppel and J. J. Thiessen. Liquid chromatographic analysis of sulfaquinoxaline and its application to pharmacokinetic studies in rabbits. *J. Pharm. Sci.* **73**:1635–1638 (1984).
28. T. B. Vree, J. J. Reekers-Ketting, C. A. Hekster, and J. F. M. Nouws. Acetylation and deacetylation of sulphonamides in dogs. *J. Vet. Pharmacol. Ther.* **6**:153–156 (1983).
29. P. F. Coccia and W. W. Westerfeld. The metabolism of chlorpromazine by liver microsomal enzyme system. *J. Pharmacol. Exp. Ther.* **157**:446–458 (1967).
30. T. J. Jaworski, E. M. Hawes, G. Mckay, and K. K. Midha. The metabolism of chlorpromazine N-oxide in man and dog. *Xenobiotica* **20**:107–115 (1990).
31. I. Bekersky, W. A. Colburn, L. Fishman, and S. A. Kaplan. Metabolism of salicylic acid in the isolated perfused rat kidney. *Drug Metab. Dispos.* **8**:319–324 (1980).
32. J. Mann and E. Gurpide. Generalized rates of transfer in open systems of pool in the steady state. *J. Clin. Endocr.* **26**:1346–1354 (1966).
33. J. J. DiStefano. Concepts, properties, measurement, and computation of clearance rates of hormones and other substances in biological samples. *Ann. Biomed. Eng.* **4**:302–319 (1976).
34. J. H. Oppenheimer and E. Gurpide. Quantitation of the production, distribution, and interconversion of hormones. In L. J. Degroot (ed.), *Endocrinology*, Vol. 3, Grune and Stratton, New York, 1979, pp. 2029–2036.
35. J. G. Wagner, A. R. DiSanto, W. R. Gillespie, and K. S. Albert. Reversible metabolism and pharmacokinetics: Application to prednisone and prednisolone. *Res. Commun. Chem. Path.* **32**:387–405 (1981).
36. A. Rescigno and E. Gurpide. Estimation of average times of residence, recycle, and interconversion of blood-borne compounds. *J. Clin. Endocrinol. Metab.* **36**:263–276 (1973).
37. H. Cheng and W. J. Jusko. Mean interconversion times and distribution rate parameters for drugs undergoing reversible metabolism. *Pharmaceut. Res.* **7**:1003–1010 (1990).
38. S. S. Hwang, K. C. Kwan, and K. S. Albert. A linear model of reversible metabolism and its application to bioavailability assessment. *J. Pharmacokin. Biopharm.* **9**:693–709 (1981).
39. S. S. Hwang and W. F. Bayne. General method for assessing bioavailability of drugs undergoing reversible metabolism in a linear system. *J. Pharmaceut. Sci.* **75**:820–821 (1986).
40. H. Cheng and W. J. Jusko. Constant-rate intravenous infusion methods for estimating steady-state volumes of distribution and mean residence times in the body for drugs undergoing reversible metabolism. *Pharmaceut. Res.* **7**:628–632 (1990).
41. L. Aarons. Mean residence time for drugs subject to reversible metabolism. *J. Pharm. Pharmacol.* **39**:565–567 (1987).
42. H. Cheng and W. J. Jusko. Mean residence times of multicompartmental drugs undergoing reversible metabolism. *Pharmaceut. Res.* **7**:104–108 (1990).
43. H. Cheng and W. J. Jusko. Mean residence times and distribution volumes for drugs undergoing linear reversible metabolism and tissue distribution and linear or nonlinear elimination from the central compartments. *Pharmaceut. Res.* **8**:508-51 (1991).

PHARMACOKINETIC MODELS FOR ANTICANCER AND ANTIVIRAL DRUGS FOLLOWING ADMINISTRATION AS NOVEL DRUG DELIVERY SYSTEMS

James M. Gallo

Department of Pharmaceutics
University of Georgia

INTRODUCTION

Physiologically-based pharmacokinetic (PB–PK) models are a valuable means to evaluate drug distribution in tissues. PB–PK models, with its origins in chemical engineering [1], provide a unified and comprehensive account of drug transport by utilization of differential mass balance equations. The equations and associated parameters, such as organ blood flows and tissue to blood partition coefficients, yield a framework to represent mechanistic information on drug transport. PB–PK models offer advantages of providing predicted tissue drug concentrations as a function of time for different experimental conditions, and by model scale-up for humans. PB–PK models are distinct from other pharmacokinetic data analysis methods in being able to predict individual tissue drug concentrations, and to investigate alterations in physiological parameters on drug disposition. The advantages of PB–PK modeling are counterbalanced by variable parameter estimation methods, lack of uniform model discrimination methods and requirement of large data bases for either model development or validation.

PB–PK models have been developed for numerous compounds in a variety of therapeutic classes, although anticancer drugs have been the focus of most models [2]. Models have been proposed in numerous animal species, yet unfortunately, few have been scaled to man. The paucity of animal-based models that have been extrapolated to man may reflect the uncertainty in the amount of tissue drug concentration-time data needed to develop scaling or allometric relationships, and if scaling of parameters, such as partition coefficients is needed in all cases. In many cases, to circumvent this problem, investigators assume that thermodynamic and membrane flux parameters are equivalent in animals and in man.

The advent of novel drug delivery systems and their application to virtually all diseases require pharmacokineticists to rethink what data analysis and modeling approaches will be most useful. For systems in which therapeutic agents are associated with a carrier, consideration of disposition processes for both the active agent and carrier will be requisite for comprehensive pharmacokinetic models. Since many carrier systems are designed to undergo specific cellular uptake processes, and to release the chemical agent in a prescribed manner, the PB–PK modeling approach is a logical means to incorporate these and other factors into a model. Of course, as with studies on disposition of single drug entities, the goal of the investigation

Advanced Methods of Pharmacokinetic and Pharmacodynamic Systems Analysis
Edited by D'Argenio, Plenum Press, New York, 1991

should dictate the pharmacokinetic data analysis or modeling methods. In the area of targeted drug delivery, the objective of predicting target and other organ drug concentrations is foremost. Model predictions can be used to evaluate organ drug delivery under numerous experimental conditions, and thus, provide feedback to design future systems. One example of this latter idea would be to correlate membrane flux parameters, contained in the PB–PK models, and chemical structure for a series of surface-coated liposomes. The functional relationship could then be used to predict membrane flux for a chemical considered as a liposome coat.

Comprehensive pharmacokinetic models for novel drug delivery systems are in their infancy. The first such PB–PK model for drug distribution that incorporated differential equations for both free and entrapped drug was published only recently [3]. The same limitations, such as parameter estimation and collection of large data sets, that have hindered PB–PK approaches for simple drugs, are further magnified when one considers this approach for a novel drug delivery system.

The objectives of this communication are to demonstrate that PB–PK approaches for drug delivery systems are manageable, and at times systematic. Recommendations for future PB–PK modeling efforts will be directed to problems uncovered in the current investigations.

PARAMETER ESTIMATION

PB–PK models can vary in complexity from blood flow-limited to distributed parameter models in which both temporal and spatial dependencies of drug concentrations are determined. The vast majority of models utilize either blood flow-limited or membrane-limited organ representations, and require estimation of organ blood flows, tissue volumes, drug clearances, partition coefficients, and for membrane-limited cases, membrane flux parameters. Of these parameters, organ blood flow and tissue volumes are normally obtained from the literature, although large differences in reported values are apparent. From most investigator's perspective, the primary limitation to initiation of a PB–PK model is obtaining values for partition coefficients and membrane fluxes, typically embodied in a mass transfer coefficient.

Both *in vitro* and *in vivo* methods have been applied to estimate tissue-to-plasma partition coefficients [4–6]. *In vitro* methods are appealing in that the *a priori* flavor of the modeling effort is retained. Simulations can be attained prior to collecting animal data, and if the data is used only for model validation, less data may be sufficient. *In vivo* methods may be advantageous in terms of practicality. The animal data will serve as a means to estimate parameters and to validate the models. The time and effort required for the *in vitro* experiments could be saved.

Techniques to calculate mass transfer coefficients for PB–PK models have been somewhat obscure, in part due to less use of membrane-limited organ structures. *In vitro* cellular transport studies are not routinely done, and justifiably so, based on potential discrepancies in applying an *in vitro* parameter based on a single cell type to an organ composed of multiple cell types. *In vitro* cellular studies may be attractive when specific cell types are represented in the model.

For *in vivo* methods, and a non-eliminating organ as illustrated in Fig. 1, the mass transfer coefficient, h, can be estimated by [7]:

$$h = \left(\frac{V_T(AUMC)_T - (V_E + RV_I)(AUMC)_p}{V_I^2 R^2 (AUC)_P} - \frac{V_E^2 + 2V_E V_I R + V_I^2 R^2}{Q V_I^2 R^2} \right)^{-1} \tag{1}$$

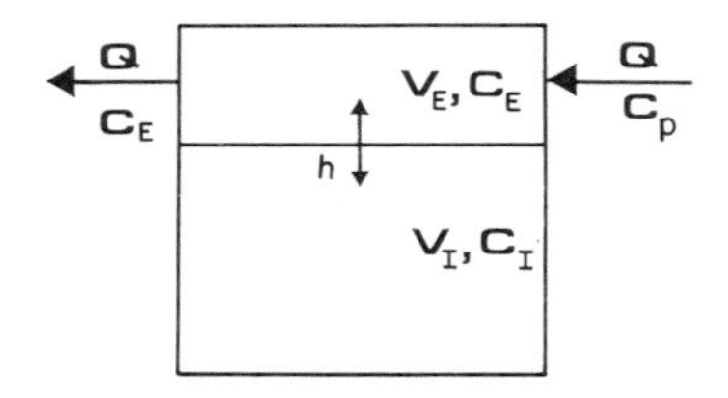

Fig. 1. Representation of a non-eliminating membrane-limited organ.

where

$AUMC_{T,P}$ = area under the first moment curve for tissue (T) or plasma (P),

R = partition coefficient,

V_E = volume of the extracelluar compartment,

V_I = volume of the intracelluar compartment,

V_T = total tissue volume, and

Q = organ flood flow.

The areas needed for Eq. (1) can be obtained from the observed concentration-time data by numerical integration techniques. The partition coefficient, R, can also be obtained from the observed data by the area method [6]. Total tissue volume (V_T) and organ blood flow (Q) can be determined experimentally or obtained from the literature. Partitioning of total tissue volumes into intracellular and extracellular fractions can be done by literature methods [1]. Estimation of h by Eq. (1) is referred to as the moment method.

The accuracy and precision of the moment method was tested using a membrane-limited model for streptozoticin in mice [8]. The reported model parameters were used to generate original streptozoticin concentration-time data. By Monte Carlo simulation, 1000 error streptozoticin concentration-time data sets were produced at maxima of 5, 10, 20 and 40 percent error. Mass transfer coefficients were calculated by Eq. (1) for each error data set for the intestine, pancreas and spleen compartments, and compared to the original parameters (see Table I). It was concluded that the moment method was a robust parameter estimation technique, and only at high percentage errors did estimations deviate from the original value (see spleen, Table I). The moment method was also compared to the forcing function method, another technique based on *in vivo* tissue concentration data [7].

The above parameter estimation methods are specific to drug disposition, following administration of drug or following carrier administration once drug is released from the carrier. Methods to estimate parameters for carrier disposition have not been formally developed. The types of parameters introduced by carrier-based systems include mass transfer coefficients for carriers, and drug release rate constants from the carrier. Empirical [3] and forcing function techniques [9] have been used in these cases, however, future work in this area would be beneficial.

Table I: Mass Transfer Coefficients Obtained from 1000 Streptozoticin Data Sets at each Error Level *

Tissue	Minimum Percentage Error	Mass Transfer Coefficients (ml.min)		
		Mean	+ SD	% Bias
Intestine	5	0.6196×10^{-01}	0.9278×10^{-03}	-3.34
	10	0.6196×10^{-01}	0.1856×10^{-02}	-3.34
	20	0.6196×10^{-01}	0.3712×10^{-02}	-3.34
	40	0.6205×10^{-01}	0.7438×10^{-02}	-3.20
Pancreas	5	0.2681×10^{-03}	0.3234×10^{-05}	-10.63
	10	0.2680×10^{-03}	0.6469×10^{-05}	-10.65
	20	0.2679×10^{-03}	0.1294×10^{-04}	-10.69
	40	0.2677×10^{-03}	0.2591×10^{-04}	-10.78
Spleen	5	0.5974×10^{-03}	0.2799×10^{-04}	19.48
	10	0.6003×10^{-03}	0.5681×10^{-04}	20.07
	20	0.6135×10^{-03}	0.1208×10^{-03}	22.70
	40	0.6873×10^{-03}	0.3483×10^{-03}	37.46

*Original mass transfer coefficients were 0.0641 ml/min, 0.0003 ml/min and 0.0005 ml/min for the intestine, pancreas and spleen, respectively.

PHYSIOLOGICAL PHARMACOKINETIC MODELS OF ANTICANCER DRUGS

A PB–PK model for adriamycin following administration as a solution and magnetic albumin microspheres has been reported recently [3]. The steps in producing the models were to first characterize the disposition of adriamycin administered as a solution (see Fig. 2), and then superimpose onto this a microsphere transport model (see Fig. 3). Thus, total adriamycin concentrations in each organ assumed to contain adriamycin and microspheres were represented as:

$$\left(\frac{dC}{dt}\right)_{total} = \left(\frac{dC}{dt}\right)_S + \left(\frac{dC}{dt}\right)_{MS} \qquad (2)$$

where dC/dt is the rate of change of adriamycin concentration in solution form (S) or associated with the microspheres (MS). The equations representing $(dC/dt)_S$ are exactly the ones used for solution administration of adriamycin. The rate processes, $(dC/dt)_S$ and $(dC/dt)_{MS}$ were linked by first-order release rate constants for adriamycin from the microspheres. Model parameters consistent with the model illustrated in Fig. 2 were estimated from methods indicated above, and include the area [6] and moment methods [7]. This model and associated parameters were not altered in the model describing adriamycin kinetics following microsphere administration. New parameters and mass balance equations, consistent with Fig. 3, were added to the model depicted in Fig. 1. These new parameters, such as release rate constants, were obtained by trial and error based on agreement between observed and predicted adriamycin concentrations. Physiological homogeneity of the parameters was maintained.

The model presented in Figs. 2 and 3 were developed based on data collected following administration of 2 mg/kg of adriamycin. Subsequently, 0.04 mg/kg and 0.4 mg/kg of adriamycin as magnetic albumin microspheres were studied under the same conditions. The same model used for the 2 mg/kg dose study was applied to

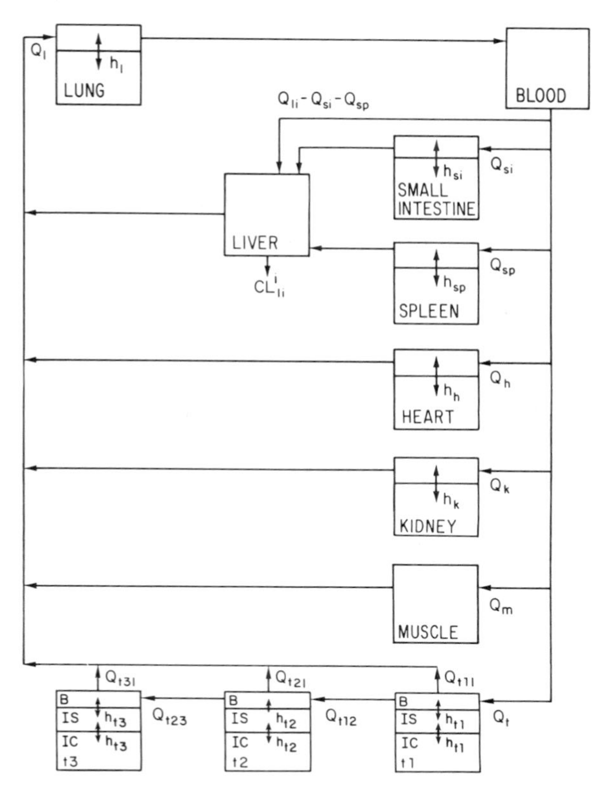

Fig. 2. Physiological pharmacokinetic model for adriamycin in the rat illustrating its disposition in "free" form. The model was developed based on data collected following administration of adriamycin in solution. B = blood, IS = interstitial, IC = intracellular, t1 = tail compartment 1, t2 = tail compartment 2, t3 = tail compartment 3, *Q*'s represent organ blood flows, *h*'s represent mass transfer coefficients, CL_{li} = intrinsic liver clearance.

the lower dose studies by simply changing the administered dose. Representative adriamycin concentrations from the three dosage levels are shown in Figs. 4, 5 and 6. The target site (t2), a mid-segment of the rat tail, adriamycin concentrations (see Fig. 2) were overpredicted until 24 hr at the 2 mg/kg dose, whereas model predictions for the two lower doses either agree or underpredict the actual data. One could conclude that the same model, which assumed linear drug and microsphere transport, is not adequate to characterize adriamycin's disposition over this range of doses. Further examination of the dose-normalized *AUC* for adriamycin at t2 revealed a nonlinearity. At the 0.04mg/kg and 0.4 mg/kg doses, the dose-normalized *AUC*s were equivalent, yet decreased by 60% at the high dose. The dose-normalized *AUC*s at t2 following solution administration of 0.12 mg/kg and 2 mg/kg of adriamycin did not show this trend, in fact, the dose-normalized *AUC* increased with an increase in dose. Therefore, as the model (see Fig. 4a) and data indicate, there appears to be a saturable microsphere transport process at the target site that resulted in lower than expected adriamycin concentrations at the 2 mg/kg dose.

Analysis of observed and predicted liver adriamycin concentrations (see Fig. 5) indicate a similar phenomenon as the target site. Predictions at the two lower doses better agree with observed data than those obtained at the high dose, where overpredictions are seen until late time points. Again, this suggests saturable microsphere

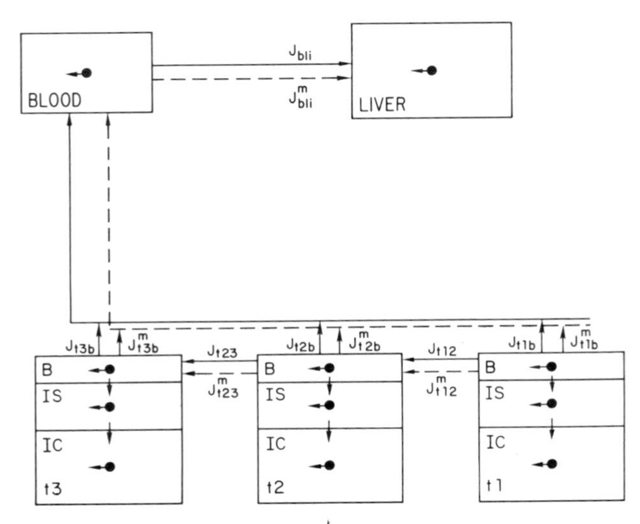

Fig. 3. Physiological pharmacokinetic model for adriamycin associated with magnetic albumin microspheres in the rat. J's represent intercompartmental transfer rates of microspheres containing adriamycin in the presence (m) and absence of the magnetic field. ↓ indicates microsphere mass transfer coefficient, ↔• indicates microsphere and associated adriamycin release rate constant. Other abbreviations defined in Fig. 1.

transport processes may be involved at the liver. The blood flow-limited model for the liver ignores membrane resistances, and thus, a membrane-limited model for microspheres in the liver may be appropriate.

Finally, predicted and observed adriamycin concentrations for the heart, an organ of significant toxicological interest, are shown in Fig. 6. Overall, predicted values agree with the observed data, and evidence for nonlinear processes for free drug transport (microspheres were assumed not to distribute to the heart) are absent. Except for the aberrant observed adriamycin concentrations at 2 hr for the 0.4 mg/kg dose, all model predicted concentrations were within 0.5 μg/ml or less. The magnitude of this error would probably not lead to errors in interpretation of adriamycin disposition.

Investigations into nonlinear microsphere transport models as suggested above would best be done by obtaining measurements, such as radioactivity, of microsphere tissue concentrations.

PHARMACOKINETIC MODELS FOR ANTIVIRAL DRUGS

Few PB–PK models for antiviral compounds have been developed. However, with the advent of AIDS, and the resurgence of antiviral therapy, modeling efforts will undoubtedly increase. Further, since the central nervous system is a target site for anti-HIV therapy, development of predictive pharmacokinetic models will be useful to evaluate targeted drug delivery systems, and dosage regimens.

Hybrid pharmacokinetic models have been developed to characterize zidovudine (AZT), and azidouridine (AZdU, AZddU) brain disposition following iv administration as parent and prodrugs [9]. Hybrid models combine features of both classical compartmental and physiological pharmacokinetic models. In contrast to global PB–PK models (see Fig. 2), hybrid models normally focus on a single organ

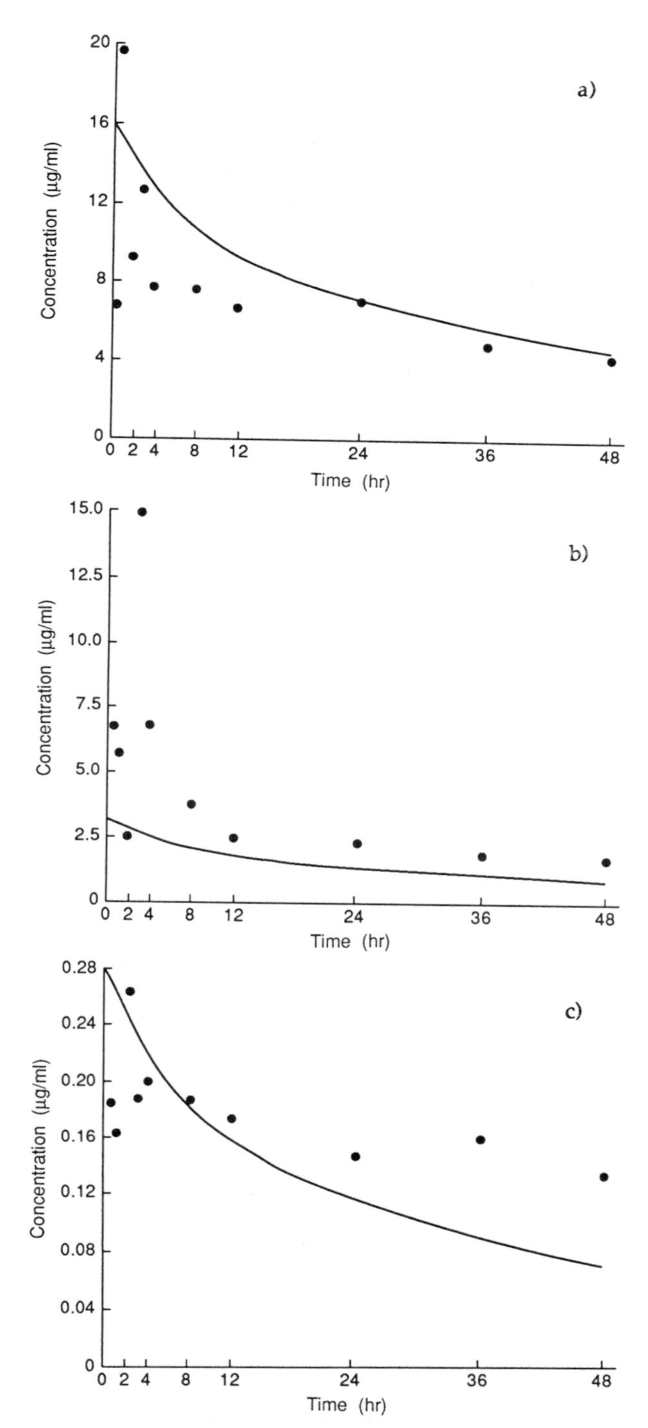

Fig. 4. Observed (•) and model predicted (—) t2 adriamycin concentrations following administration of magnetic albumin microspheres containing adriamycin to the rat at dose of a) 2.0 mg/kg, b) 0.4 mg/kg, c) 0.04 mg/kg. A magnetic field of 8000 Gauss was applied to t2 for 30 min following microsphere administration in all studies.

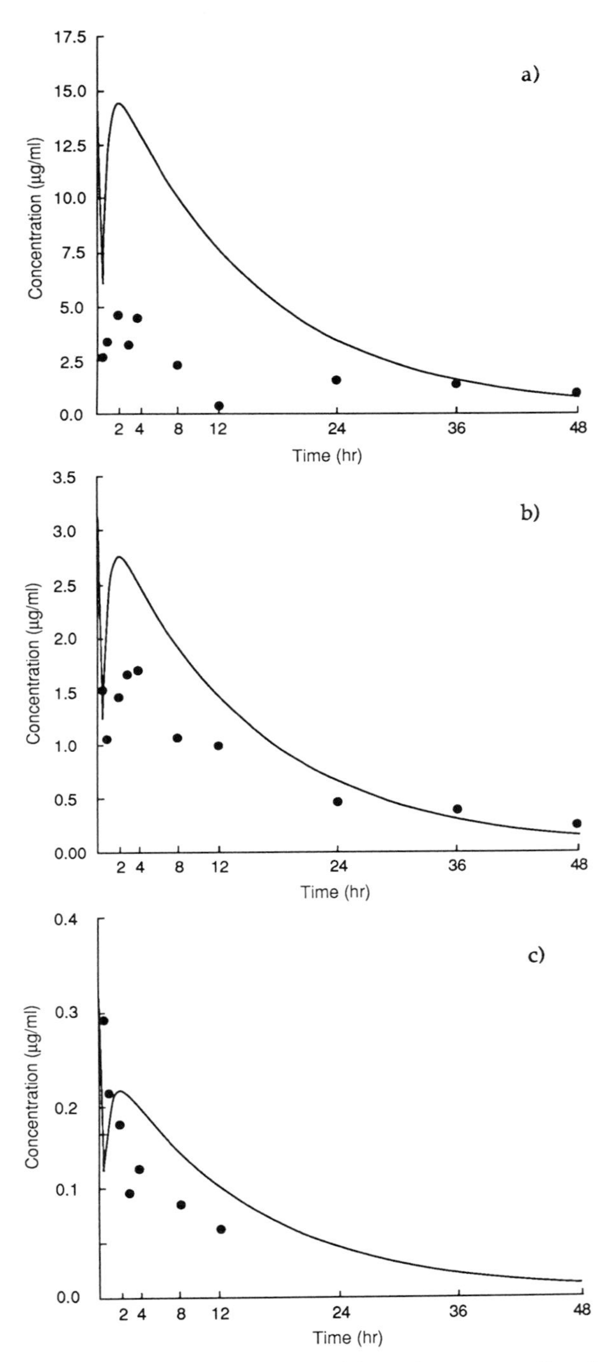

Fig. 5. Observed (•) and model predicted (—) liver adriamycin concentrations following administration of magnetic albumin microspheres containing adriamycin to the rat at dose of a) 2.0 mg/kg, b) 0.4 mg/kg, c) 0.04 mg/kg. A magnetic field of 8000 Gauss was applied to t2 for 30 min following microsphere administration in all studies.

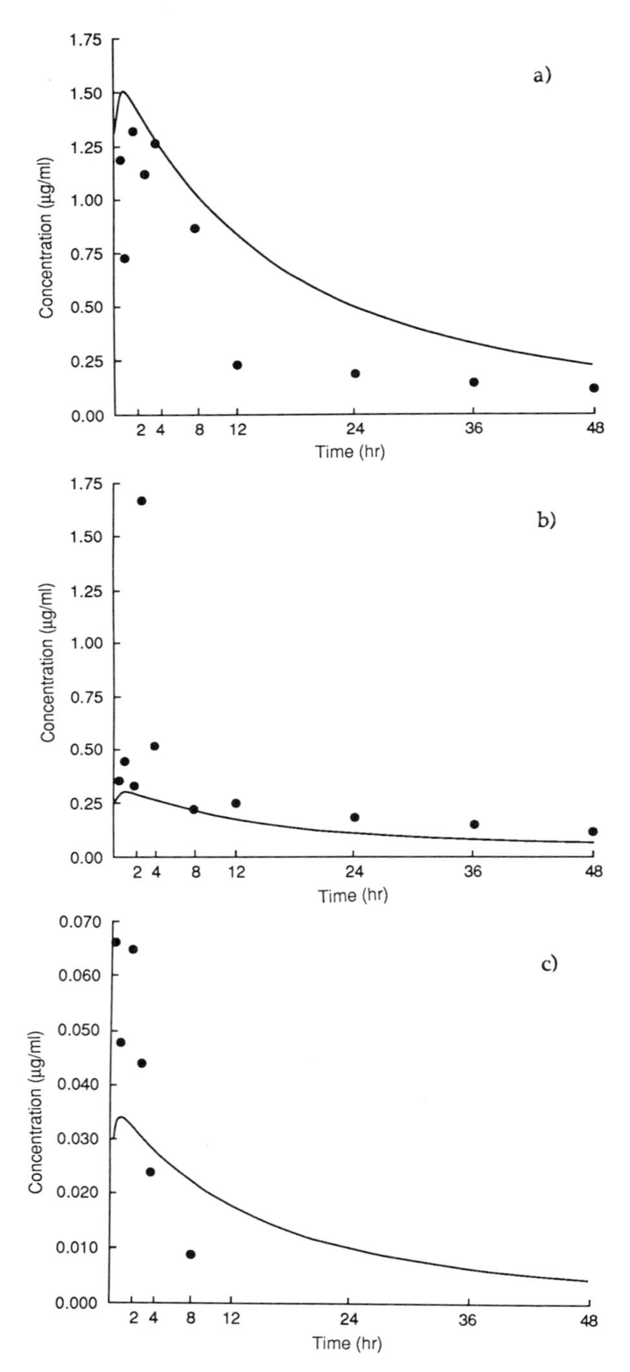

Fig. 6. Observed (•) and model predicted (—) heart adriamycin concentrations following administration of magnetic albumin microspheres containing adriamycin to the rat at dose of a) 2.0 mg/kg, b) 0.4 mg/kg, c) 0.04 mg/kg. A magnetic field of 8000 Gauss was applied to t2 for 30 min following microsphere administration in all studies.

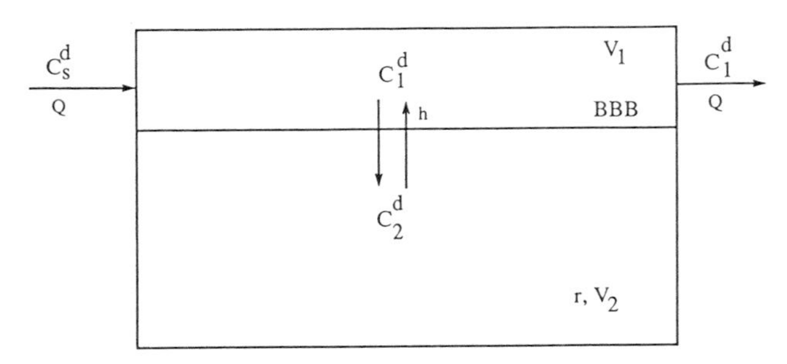

Fig. 7. Hybrid pharmacokinetic model characterizing AZT and AZddU disposition in the mouse brain.

of interest. The organ representation is physiologic as in a global model, however, drug concentrations (typically plasma) entering the organ are expressed as polyexponential equations obtained by nonlinear regression data fitting procedures. Less parameters and experimental data are required for a hybrid model compared to a global PB–PK model, yet the predictive nature of global models is still preserved.

Figures 7 and 8 illustrate the models for anti-HIV nucleosides and their dihydropyridine prodrugs in mice. In each case, the brain is comprised of two compartments, vascular and extravascular, separated by the blood-brain barrier (BBB). Transport of AZT, AZddU and of the dihydropyridine prodrugs (AZT–DHP, AZddU–DHP) is considered to be a passive or first-order process, as is the metabolism of the prodrugs to the quaternary salt derivatives (AZT–QS, AZddU–QS) and to AZT and AZddU. The quaternary salt metabolites were assumed not to cross the BBB.

Model parameters were estimated by a variety of methods, and details are provided elsewhere [9]. Methods included literature values (Q, V_1, V_2), *in vitro* techniques, (r_{pd}, K_i), forcing function (h, r), and empirical (r_{pd}, h_{pd}, K_i) methods. Serum concentrations for each species entering the brain compartment were represented by polyexponential equations obtained by nonlinear regression analysis. These equations were substituted into the differential mass balance equations describing the

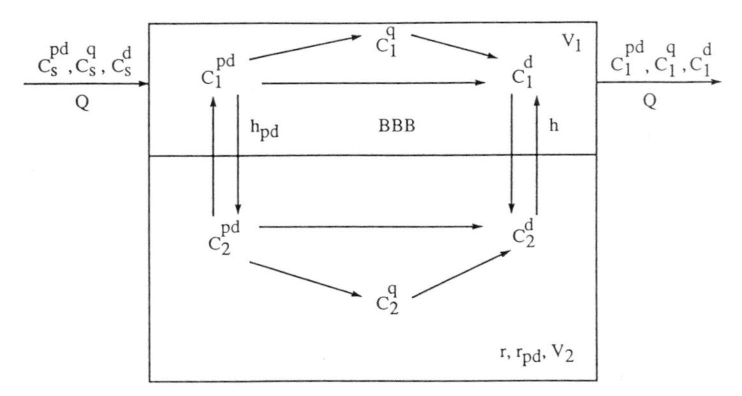

Fig. 8. Hybrid pharmacokinetic model characterizing AZT–DHP, AZddU–DHP and their metabolites disposition in the mouse brain.

models in Figures 7 and 8. The mass balance equations were solved numerically, and the resultant concentrations of each species in compartments 1 and 2 were volume averaged to provide predicted brain concentrations. Figures 9–12 show the observed and predicted concentrations for the four studies. Model predicted AZT (Fig. 9) and AZddU (Fig. 10) brain concentrations agree well with the observed brain concentrations. The solid-line model predicted concentrations in Figs. 11 and 12 indicate reasonable agreement with the data obtained following AZT–DHP and AZddU–DHP administration, respectively. The dotted-line predictions, which utilized metabolic constants (K_i) determined from *in vitro* studies conducted in mouse brain homogenate, are more divergent, and illustrate potential problems in obtaining parameters exclusively from *in vitro* studies. Another potential source of error in the prodrug models is the forcing functions or exponential equations describing C_s. Due to the rapid conversions of the prodrugs in serum, the exponential equations are based on a small number of data points. Nonetheless, predicted brain AZT, and particularly AZddU concentration simulate the observed data.

Another example of a hybrid model for AZT brain disposition in rabbits is illustrated in Fig. 13. This model was put forth by Spector *et al.* to predict mannitol and ascorbic acid cerebrospinal fluid (CSF) and brain concentrations in rabbits [10]. The model is general in that linear and nonlinear solute transport processes are represented between the blood or plasma, extracellular space (ECS)–CSF, and brain parenchyma compartments.

New Zealand male rabbits (bw = 2 kg) were administered 17.2 mg/kg AZT iv, and euthanized by sodium pentobarbitol for up to 3 hr after administration. At sacrifice, blood, CSF and brain were collected and analyzed for AZT by HPLC. Parameters used for the simulation were obtained from the literature, by empirical methods, and by nonlinear regression to obtain parameters for the biexponential equation that described blood concentrations. Parameters values are listed in Table II. Observed and model predicted CSF and total brain AZT concentrations are presented in Fig. 14. The predictions agree quite closely to the data, and demonstrate that the hybrid modeling approach is useful for sophisticated brain transport models. Extension of this model to an AZT–DHP study in rabbits was not possible since AZT–DHP blood concentrations were below assay sensitivity, thus preventing a forcing function to be attained.

CONCLUSIONS

PB–PK models offer a powerful tool to predict organ drug concentrations following administration of drug alone or as a novel drug delivery system. Through the modeling efforts presented here, further work in the areas of parameter estimation, model discrimination and study designs are recommended. Parameter estimation by empirical means may be overused, whereas *in vitro*, forcing function, and area/moment methods offer standardized approaches. Reliance on graphical comparisons of observed and predicted values to distinguish between models is tedious and subjective. Promotion of quantitative goodness-of-fit tests are encouraged [11]. Scale-up of animal-based PB–PK models to humans is infrequent, possibly due to the preception of requiring large data bases. Statistically-based study designs across multiple species may help to alleviate this problem. Advances in these areas will lead to standardized quantitative means to formulate and validate PB–PK models, and promote their greater use.

Table II: Parameters Used in Hybrid Model Describing AZT Brain Disposition Following Administration of 17.2 mg/kg of AZT iv to Rabbits

Parameter	Description	Value
A_1, A_2	y-axis intercepts for equation describing AZT in blood	29.66, 2.552 μg/ml
a_1, a_2	disposition rate constants for equation describing AZT in blood	0.1193, 0.01454 min^{-1}
V_{CSF}	volume of ECS–CSF compartment	3.0 ml
V_{br}	volume of intracellular brain compartment	6.0 ml
V_{m1}	maximum transport rate from blood to ECS–CSF	1.6 μg/min
K_{m1}	Michaelis constant for blood to ECS–CSF	5.0 μg/ml
K_1	diffusion constant from blood to ECS–CSF	0.004 ml/min
K_2	diffusion constant to and from ECS–CSF and intracellular brain compartment	1.0 ml/min
K_3	diffusion constant from ECS–CSF to blood	0.5 ml/min
K_4	bulk flow constant from ECS–CSF to blood	0.009 ml/min
R_{CSF}	partition coefficient for ECS–CSF compartment	1.0
R_{br}	partition coefficient for intracellular brain compartment	1.0

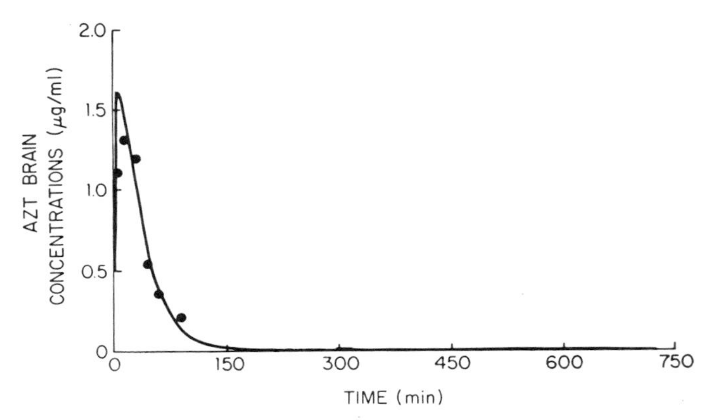

Fig. 9. Observed (•) and model predicted (—) AZT brain concentration following administration of 50 mg/kg AZT iv in mice.

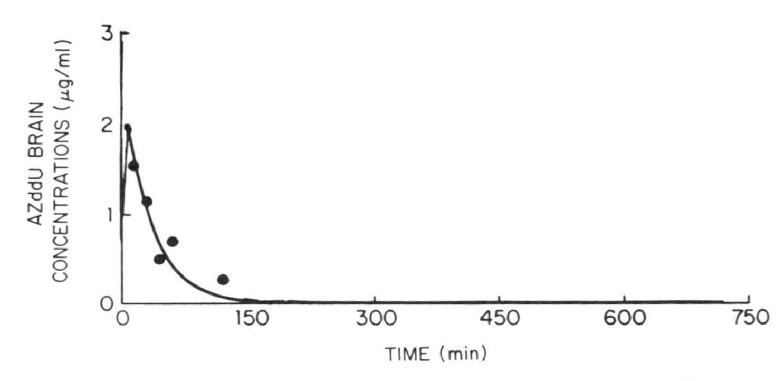

Fig. 10. Observed (•) and model predicted (—) AZddU brain concentration following administration of 50 mg/kg AZddU iv in mice.

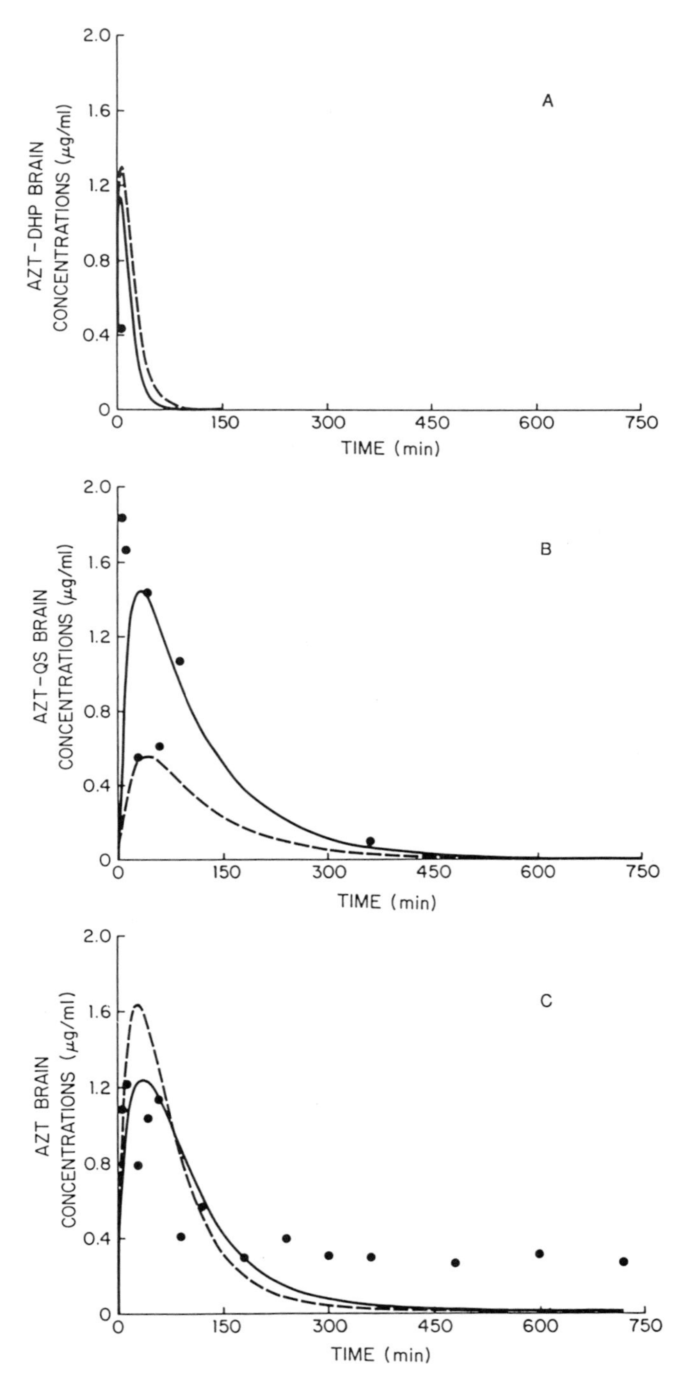

Fig. 11. Observed (•) and model predicted brain concentrations for A) AZT–DHP, B) AZT–QS, C) AZT. Dashed line (– – –) model predictions based on *in vitro* determination of metabolic rate constants, whereas solid-line (—) predictions based on empirical adjustment of metabolic rate constants determined *in vitro*. Data obtained following administration of 72.7 mg/kg (equimolar to 50 mg/kg of AZT) AZT–DHP iv to mice.

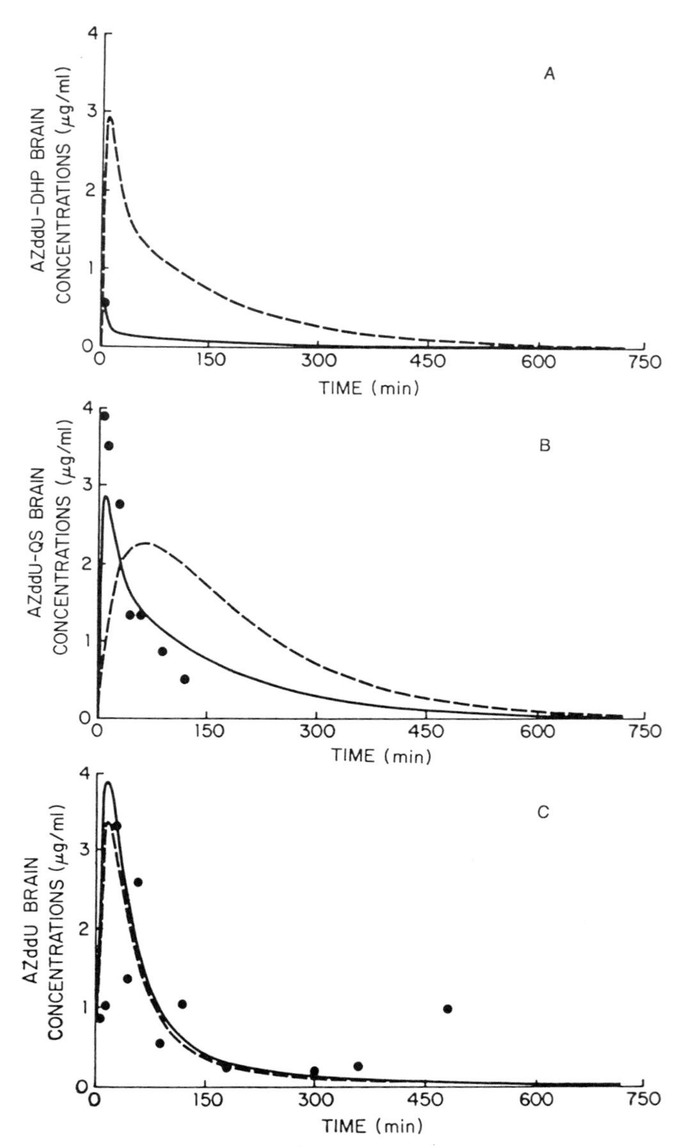

Fig. 12. Observed (•) and model predicted brain concentrations for A) AZddU–DHP, B) AZddU–QS, C) AZddU. Dashed line (– – –) model predictions based on *in vitro* determination of metabolic rate constants, whereas solid-line (—) predictions based on empirical adjustment of metabolic rate constants determined in vitro. Data obtained following administration of 73.9 mg/kg (equimolar to 50 mg/kg of AZddU) AZddU–DHP iv to mice.

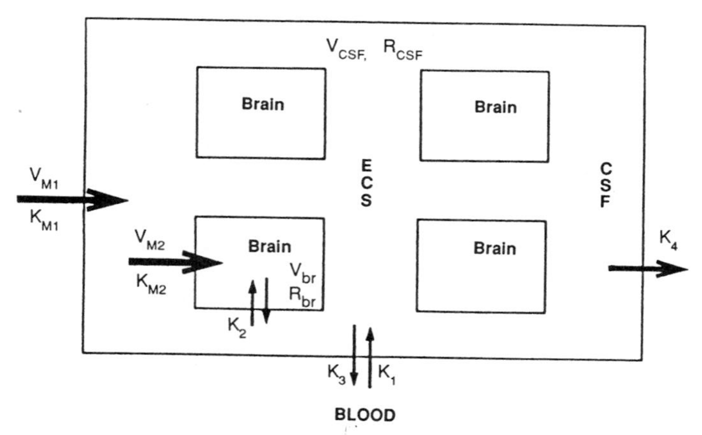

Fig. 13. Model for central nervous system drug transport applied to AZT. The linear process represented by V_{m2}. K_{m2} was found to be unnecessary in the AZT model.

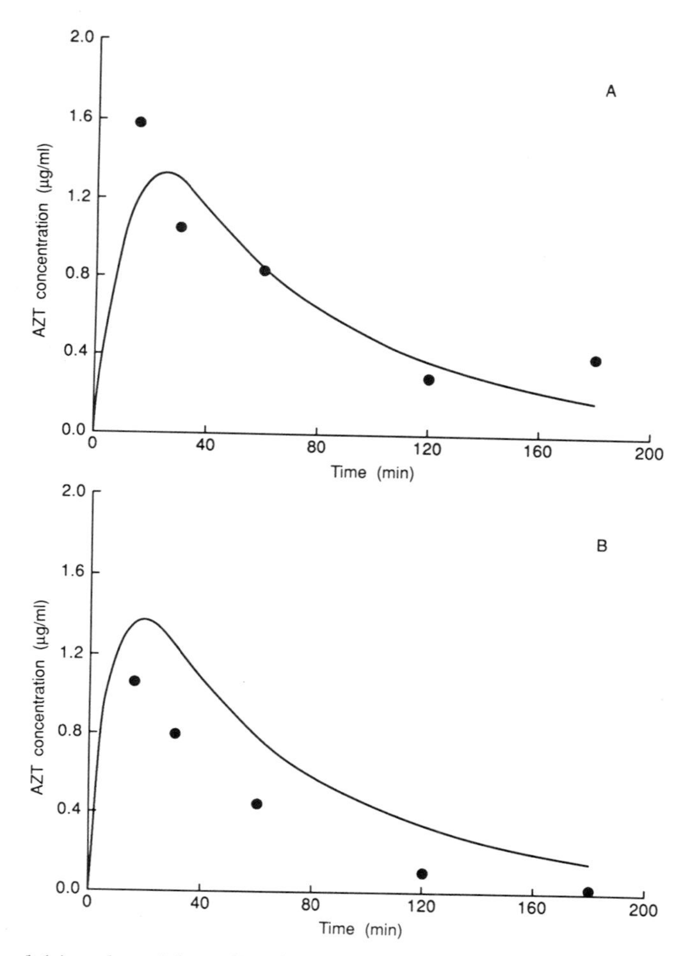

Fig. 14. Observed (•) and model predicted (—) AZT a) total brain, and b) CSF concentrations in rabbits administered 17.2 mg/kg AZT iv.

ACKNOWLEDGMENTS

The adriamycin concentration–time data was kindly provided by Drs. Tak Hung and P. K. Gupta. The rabbit zidovudine concentration–time data was kindly provided by Drs. Marcus Brewster and Wes Anderson.

REFERENCES

1. K. B. Bischoff and R. G. Brown. Drug distribution in mammals. *Chem. Eng. Prog.* **62**:33–45 (1966).
2. L. E. Gerlowski and R. K. Jain. Physiologically based pharmacokinetic modeling: Principles and applications. *J. Pharm. Sci.* **72**:1103–1127 (1983).
3. J. M. Gallo, C. T. Hung, P. K. Gupta, and D. G. Perrier. Physiological pharmacokinetic model of adriamycin delivered via magnetic albumin microspheres in the rat. *J. Pharmacokin. Biopharm.* **17**:305–326 (1989).
4. J. H. Lin, Y. Sugiyama, S. Awazu, and M. Hanano. *In vitro* and *in vivio* evaluation of the partition coefficients for physiological pharmacokinetic models. *J. Pharmacokin. Biopharm.* **10**:637–647 (1982).
5. H. S. G. Chen and J. F. Gross. Estimation of tissue-to-plasma partition coefficients used in physiological pharmacokinetic models. *J. Pharmacokin. Biopharm.* **7**:117–125 (1979).
6. J. M. Gallo, J. C. Lam, and D. G. Perrier. Area method for the estimation of partition coefficients for physiological pharmacokinetic models. *J. Pharmacokin. Biopharm.* **15**:271–280 (1987).
7. J. M. Gallo, F. C. Lam, and D. G. Perrier. Moment method for the estimation of mass transfer coefficients for physiological pharmacokinetic models. *Biopharm. Drug Dispos.* (in press).
8. J.M. Weissbrod. Comprehensive approach to whole body pharmacokinetics in mammalian systems: Applications to methotrexate, streptozotocin, and zinc. D. Engin. Sci. Thesis, Columbia University, New York, 1979.
9. J. M. Gallo, J. T. Etse, K. Doshi, F. D. Boudinot, and C. K. Chu. Hybrid pharmacokinetic models describe anti–HIV nucleoside brain disposition following parent and prodrug administration in mice. *Pharmaceut. Res.* **8**:247–253 (1991).
10. R. Spector, A. Z. Spector, and R. Snodgrass. Model for transport in the central nervous system. *Am. J. Physiol.* **232**:R73–R79 (1977).
11. J. M. Gallo, L. L. Chueng, H. J. Kim, J. V. Bruckner, and W. R. Gillespie. Physiological pharmacokinetic models for carbon tetrachloride: Use of oral absorption input functions obtained by deconvolution. Submitted for publication.

THE SIGNIFICANCE OF MARKED "UNIVERSAL" DEPENDENCE OF DRUG CONCENTRATION ON BLOOD SAMPLING SITE IN PHARMACOKINETICS AND PHARMACODYNAMICS

Win L. Chiou

Department of Pharmacodynamics
The University of Illinois at Chicago

INTRODUCTION

Until recently [1–3] it has been commonly assumed in pharmacokinetics and pharmacodynamics that "blood is blood" and blood (plasma or serum) concentrations of an endogenous or exogenous compound are practically identical, whether the blood sample is obtained from an arm artery, an arm vein, a leg vein, a pulmonary artery or a jugular vein. Such a sampling site-independent concept has apparently originated from an unrigorously tested assumption that after a bolus intravenous injection, the mixing of a substance in the entire blood circulation is extremely efficient; it was said to be complete in seconds [4,5] or in three circulatory transit times, which is about three minutes in humans [6,7] and much shorter in small animals [1]. The wide use of the plasma (blood) or central compartment concept in multicompartmental or noncompartmental analysis [8–12] in the last several decades has undoubtedly contributed to the general acceptance of the above assumption.

The main purpose of this presentation is two-fold. First, it will briefly present a theoretical discussion, coupled with examples (Figs. 1–8, see [13–19]) illustrating why and how concentrations in blood or plasma of practically all drugs may vary markedly (at least in a certain period) with the sampling sites from the time of administration or absorption until all drug molecules are eliminated from the body. Second, it will briefly review potential implications in pharmacokinetics and pharmacodynamics of such a sampling-site-dependent phenomenon. For details the reader is recommended to read a recent two-series review article by Chiou [1,2] and the cited references. It should be noted that arteriovenous difference across the liver or kidney, two well-known major eliminating organs, is not included here.

RATIONALE AND EXAMPLES OF SAMPLING–SITE DEPENDENCY IN DRUG CONCENTRATIONS

After a drug is injected or absorbed into the systemic venous blood, it will be first carried to the right heart, then through pulmonary artery and capillaries to the lungs, through the pulmonary veins to the left heart and then to the whole body except the lungs through the aorta, systemic artery and peripheral capillaries. The drug molecules, not taken-up by peripheral tissues and organs through capillary

Advanced Methods of Pharmacokinetic and Pharmacodynamic Systems Analysis
Edited by D'Argenio, Plenum Press, New York, 1991

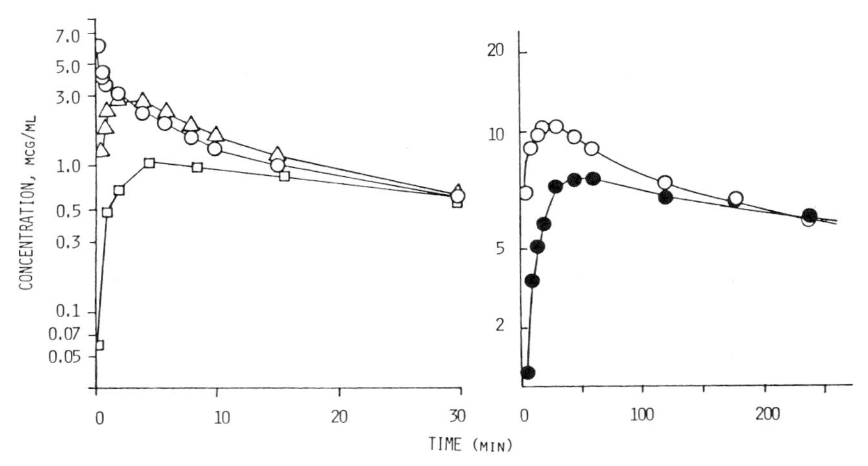

Fig. 1. Left: lidocaine plasma concentration from the pulmonary artery (–○–), brachial artery (–△–) and antecubital vein (–□–) in a subject following cuff release (45 min cuff time) after intravenous regional anesthesia with 3mg/kg of lidocaine HCl (data estimated from Tucker and Boas, 1971); Right: mean systemic arterial (–○–) and peripheral venous (–●–) lidocaine plasma levels from 10 patients after receiving extradural injection of 400 mg dose (data estimated from [13]). Marked A–V differences are expected during the terminal phase which are not shown here.

membranes, are then returning to the systemic venous blood. During the above process it is obvious that as some of the drug molecules are taken up by the lung tissue or peripheral tissues such as an arm or a leg, there will be an arteriovenous ($A - V$) difference across the tissue with arterial concentrations being higher than the venous concentrations. The extent of $A - V$ difference across a sampling tissue will depend upon the degree of uptake or "extraction" by a given tissue. The initial extraction ratio (E) during a single blood passage through a noneliminating tissue under a sink condition may be governed by the following unified organ clearance equation of Chiou [1,2,20–22]:

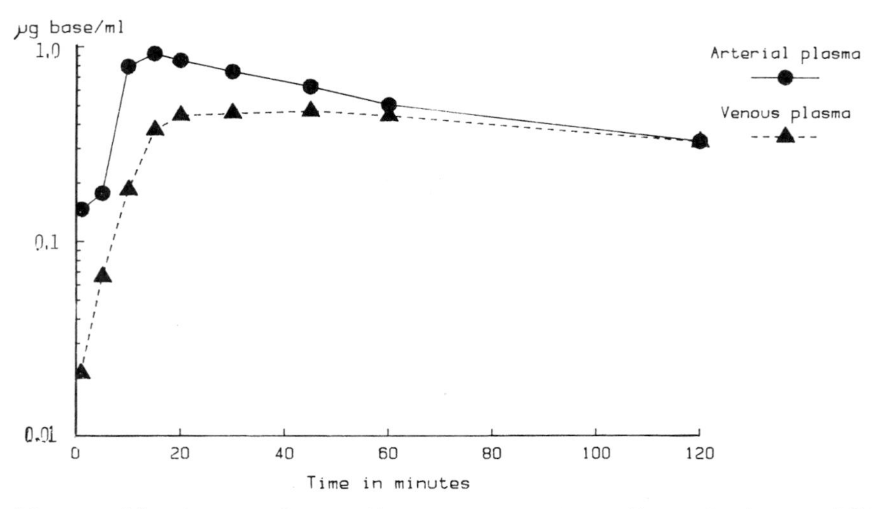

Fig. 2. Mean arterial and venous plasma etidocaine concentration profiles in 10 volunteers following a 2-min peridual administration of 20ml solution containing 1% etidocaine and 5μg/ml of adrenaline (data from [14]).

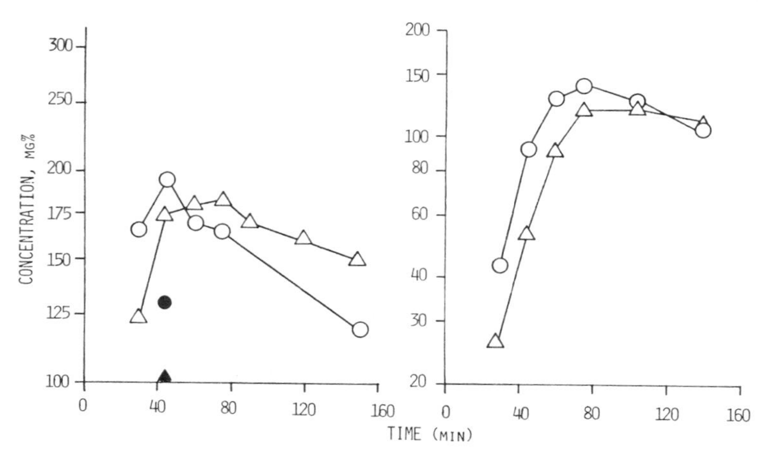

Fig. 3. Left: alcohol concentrations in blood from the cubital vein (– △ –), finger-tip capillaries (– ○ –), foot-vein (–●–) and toe tip (–▲–) in a man (subject E) after the beginning of drinking of 1.03 g per kg of alcohol in 30 minutes (systemic arterial blood concentration may be expected to peak sooner and higher than the capillary blood levels); Right: alcohol concentrations in blood from the radial artery (– ○ –) and cubital vein (– △ –) in a man (subject H) following the same dose of alcohol (data estimated from [15]). Marked $A-V$ differences are expected during the postabsorption phase which are not shown here.

$$E = H/(Q + H) \tag{1}$$

where H is the apparent uptake clearance and Q is the blood flow rate to the tissue (volume of blood flow per unit weight of sampling tissue). Molecular weight and apparent partition coefficient between tissue and plasma of drug are probably the main factors affecting the H value.

It should be noted that although muscle is a poorly perfused tissue and the mean transit time of blood through a capillary is only 1–3 seconds [1,2,23], the initial tissue uptake may be very extensive or approaching unity for many or most drugs. An E value of 0.98 will result in an about 50-fold difference in arterial and venous concentration across the tissue. It is of interest that water, glucose, sucrose and inulin molecules can diffuse back and forth through the pore of capillary membrane up to some 40, 32, 16 and 8 times, respectively, during each blood passage [23]. Lipid-soluble drugs can also penetrate through the lipoidal capillary membrane at much higher rates. The above discussion indicates that shortly after absorption or intravenous injection, practically all drugs of clinical interest will exhibit $A-V$ differences across a sampling tissue, and at each moment the degree of A–V difference may vary greatly with different sampling sites throughout the body. The maximum A/V ratios, determined at 20 seconds after intravenous injection to animals from our laboratory, were 277 for propranolol, 51 for lidocaine, 34 for procainamide, 33 for furosemide, 5.4 for theophylline and 3240 for griseofulvin [4]. The A/V ratios in humans for diazepam and phenobarbital [16] could be estimated to be approximately 100.

Theophylline has been found in healthy dogs to exhibit relatively small $A-V$ differences shortly after intravenous administration [4]. This is in sharp contrast with a recent report showing unusually large A–V differences in a patient with circulatory shock after oral administration of a sustained-release theophylline apparently for a suicidal purpose [24]. At admission, the patient's arterial plasma concentration was

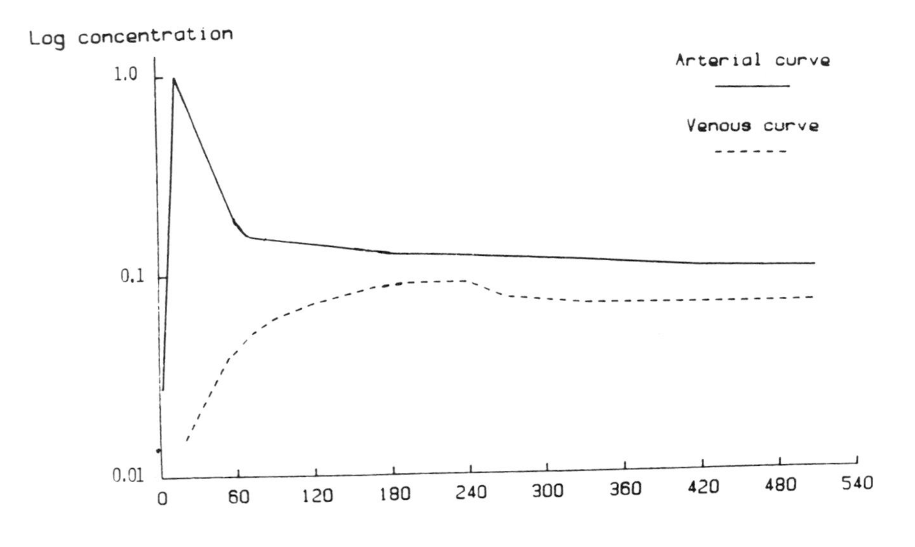

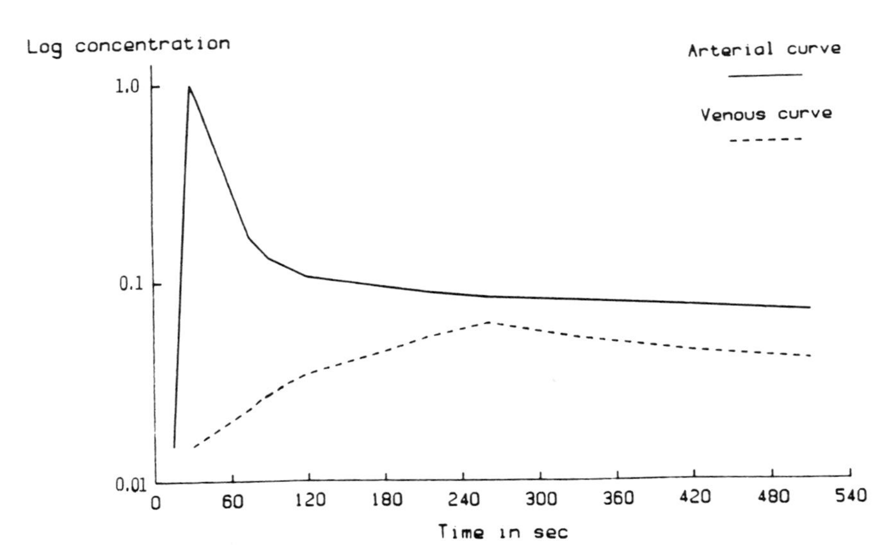

Fig. 4. Typical arterial (—) and venous (- - - -) plasma concentration profiles of labelled phenobartibal (top) and diazepam (bottom) after a rapid i.v. injection to patients (data from [16]). For the first 2 minutes samples were collected every 5 seconds, for the next 2 minutes every 10 seconds, and every minute for the next 6 minutes.

93.8 mg/L (toxic level) while the venous concentration was only 0.595 mg/L; the difference being 158-fold. At 2 and 3 hr after admission, the venous concentrations increased to 13 and 20 mg/L while arterial levels decreased to 37.3 and 35 mg/L; the A/V ratios being 2.9 and 2.75, respectively. The above seemingly unusual phenomena may be attributed to the very low Q in Eq. (1). Therefore, one should probably be very cautious in measuring venous concentrations in cases of circulatory collapse or low blood flow to a sampling tissue.

When the net tissue uptake of drug becomes zero, there will be no A–V difference across that particular tissue [1,2]. The time for this to occur in a noneliminating organ after a bolus intravenous injection may vary tremendously with sampling sites, drugs

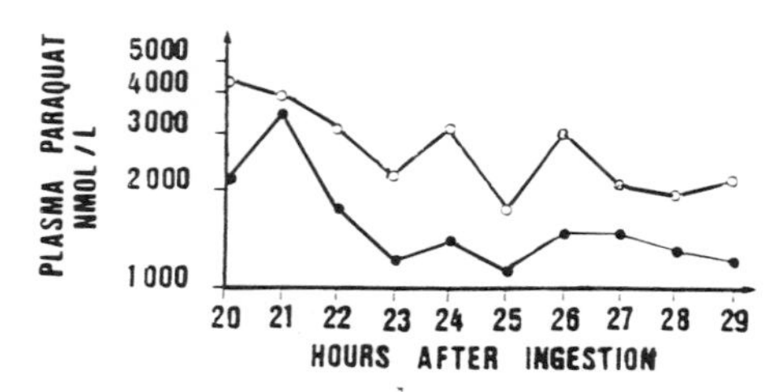

Fig. 5. Time course of pulmonary arterial (–●–) and systemic arterial (–○–) plasma concentration profiles of paraquat in a comatose patient after oral ingestion (data from [17]).

and individuals. For example, in humans this has been shown to be less than 5 minutes for insulin, and approximately 30 minutes for phenobarbital, phenytoin, clonazepam, diazepam and iodine [1]; in animals this has been shown to be about 1 minute for furosemide [1,2,4] and 30 minutes for propranolol [18] and procainamide [5], and 2 to 40 minutes for griseofulvin [1].

As the drug in the body is continuously decreasing due to elimination and uptake by other tissues, the drug taken-up or extracted earlier by a given sampled tissue will start to diffuse back into the capillaries resulting in venous concentrations being higher than arterial concentrations. The A/V ratio will then eventually reach a constant value during the terminal phase. Venous concentrations approximately two-fold higher than arterial concentrations during the terminal phase were found for procainamide, propranolol, lidocaine and trichloromonofluoromethane [1,4].

Equation (2) governing the A/V ratio during the terminal exponential phase can be derived using a simple noncompartmental mass balance principle rather than the physiologically unrealistic well-mixed compartmental model employed earlier [5]:

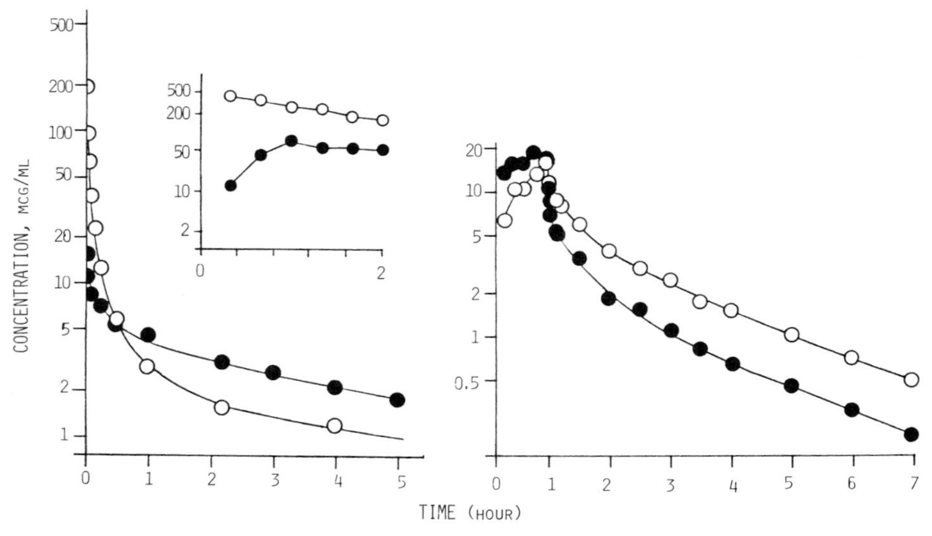

Fig. 6. Femoral arterial (–○–) and venous (–●–) plasma concentration profiles of procainamide in male rabbits. Left: for a 3.7 kg rabbit after 100 mg procainamide HCl intravenous bolus injection (upper insert showing the detailed $A-V$ profiles in the first 2 minutes following a similar study in another rabbit). Right: for a 3.2 kg rabbit following a constant intravenous infusion of procainamide HCl (244 mg/hr) for 54 minutes (data from [5]).

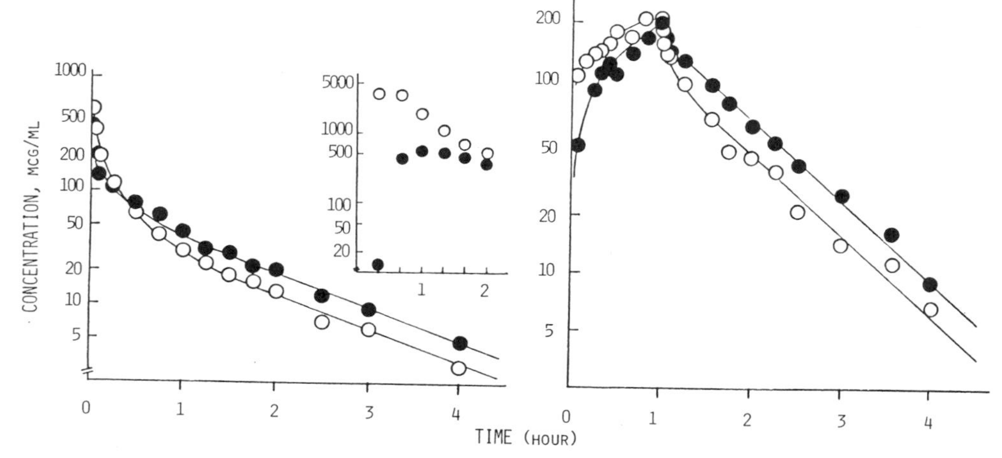

Fig. 7. Femoral arterial (–○–) and venous (–●–) plasma concentration profiles of propranolol in male mongrel dogs. Left: for a 20.3 kg dog following intravenous bolus injection of 10 mg propranolol HCl (upper insert showing detailed profiles in the first 2 minutes); Right for an 18.3 kg dog following a constant intravenous infusion of propranolol HCl (15.3 mg/hr) for 60 minutes (data from [18]).

$$A/V = 1 - (K \cdot R/Q) \tag{2}$$

where K is the first-order rate constant of the terminal phase, R is the apparent partition coefficient of the drug between the sampled tissue and the venous blood during the terminal phase and Q is the blood flow rate per unit weight of the sampled tissue. The above equation indicates that the A/V ratio during the terminal phase may vary with the sampling site in the body because of difference in R and Q. Since $K \cdot R$ may reflect the total body clearance or plasma clearance CL [1,2], the following equation (see Appendix for derivation) may be obtained:

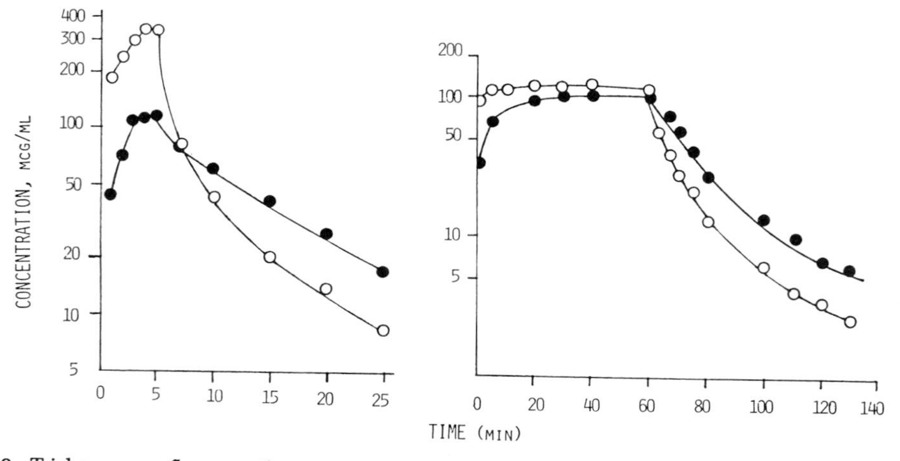

Fig. 8. Trichoromonofluoromethane concentration profiles in blood obtained from carotid artery (–○–) and femoral vein (–●–) in 2 anesthetized mongrel dogs during and following a 5-minute inhalation of 5% (v/v) in air mixture (left figure: unpublished work), and a 60-minute inhalation of 2% (v/v) in pure oxygen for 60 minutes (right figure; data from [19]).

$$(C_v - C_a)/C_v = aCL/Q \quad (3)$$

where C_v and C_a represent the venous and arterial concentrations, respectively, and a is a constant. Limited plasma $A-V$ difference and CL data for four drugs in dogs obtained from our laboratory indeed support the general trend predicted by Eq. (3) (Fig. 9). As a general rule of thumb, it may be stated that the larger the plasma clearance, the greater the $A - V$ and vice versa [1]. Obviously, a drug with a terminal half-life greater than 12–24 hr is generally expected to exhibit small or insignificant $A - V$ difference during the terminal phase in subjects with relatively normal blood flow to the sampling tissue.

The above discussion shows how the uptake and release of drug by a noneliminating sampling tissue can affect the $A-V$ difference. Although liver and kidney are two major known eliminating organs, lungs [25,26], muscle [1,27] and peripheral vascular tissues [28] can also metabolize drugs or endogenous substances thus complicating the phenomenon of $A-V$ differences [1]. An example of nitroglycerin [29] in humans is shown in Fig. 10. To date at least 45 compounds (25 in humans and 24 in animals) have been reported to exhibit various degrees of site dependency in plasma (blood or serum) concentrations [1–3,24]. It should be noted that in theory the area under the plasma-time curve from time zero to infinity (*AUC*) or the steady-state plasma level should be the same when arterial and venous samples are taken from noneliminating tissues.

SIGNIFICANCE IN PHARMACOKINETIC STUDIES

The phenomenon of potentially marked "universal" dependence of drug concentration on blood sampling site appears to have a very profound impact on many basic pharmacokinetic principles and analyses which have been commonly accepted for many decades. The implications seem too numerous to present in detail here; only some of the highlights [1,2] will be briefly mentioned below.

1. The concept of the central compartment or plasma (blood) compartment used routinely in pharmacokinetics cannot be justified fully since the plasma concentration or the plasma concentration-time profile of a drug

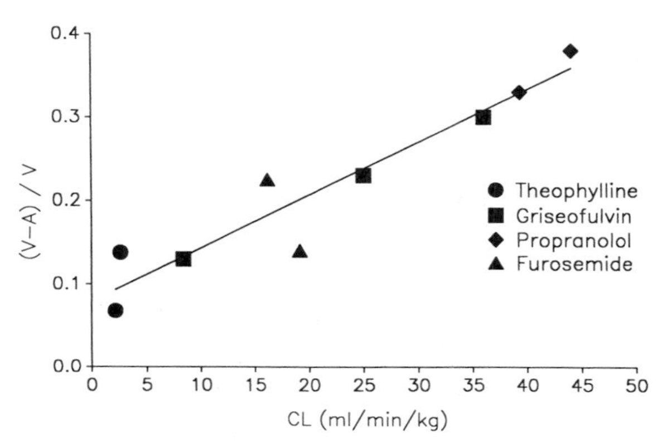

Fig. 9. Correlation between the degree of arterial-venous difference and plasma clearance of four drugs in dogs (unpublished data from W. L. Chiou).

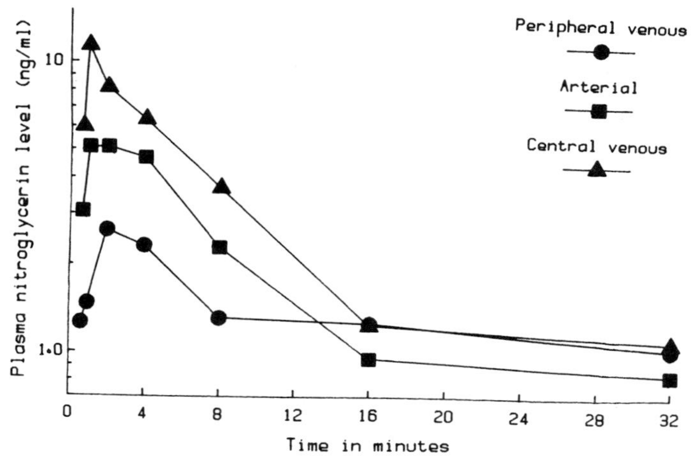

Fig. 10. Mean plasma levels of nitroglycerin following intranasal administration of an 0.8 mg dose to five patients. The concentration from the central venous blood should be equivalent to that in the pulmonary arterial blood (data from [29]).

in the body may vary greatly with the sampling site. Therefore, the concentration used to represent this compartment may be highly arbitrary depending *solely* on the site *chosen* by a *given* investigator. After a single dosing, practically all drugs are *never* homogeneously well mixed in the so-called plasma or central compartment. Therefore, it seems extremely difficult to define scientifically what is a central compartment; this is significant because pharmacokinetics has been often regarded as a mathematically sophisticated rigorous science, and ideally one should seek and be able to *validate* any basic assumptions in science.

2. It is the arterial blood that carries the drug to various parts of the body for distribution and elimination. Therefore, the "driving force" for distribution and elimination should be the drug concentration in the arterial blood, not in the venous blood as is commonly used in pharmacokinetic studies. Use of venous concentration data to estimate or study many pharmacokinetic parameters after a single dosing may be theoretically incorrect or misleading [1,2]. These include, for example, mean residence time *MRT* in the body ([30]; Table I), steady-state volume of distribution V_{ss} ([30]; Table II), renal clearance (Fig. 11), the amount of drug remaining in the body as a function of time, the time to reach a steady state, and plasma to saliva concentration ratio [1,31]. The difference in *MRT* or V_{ss} obtained from the arterial and venous data has been attributed to the mean transit time or V_{ss} of the drug in the sampling tissue [2,30].

3. As an approximation [1,2] the commonly assayed venous concentration may be more correctly regarded to reflect the concentration in the poorly perfused sampling tissue rather than in the so-called central compartment.

4. The classical instantaneous-input-to-the-central or -plasma compartment concept [32] apparently lacks any scientific support as the drug concen-

Table I. Comparison of Mean Residence Time (*MRT*) of Five Drugs in Dogs and Rabbits Estimated Based on Femoral Arterial (MRT_{sa}) or Venous (MRT_{sv}) Plasma Data [30]

Drug	Animal (min)	MRT_{sa} (min)	MRT_{sv} (%)	Difference (min)	MRT_s^*
Propranolol					
	Dog A	38.2	66.0	72.8	27.8
	Dog A	40.9	57.4	40.3	16.5
	Rabbit A	132	188	42.4	56
	Rabbit B	110	171	55.5	61
Griseofulvin					
	Dog C	56.5	71.9	27.3	15.4
	Dog D	34.3	44.5	29.7	11.2
	Dog E	94.2	104	10.8	9.8
	Rabbit C	240	313	30.4	73
Furosemide					
	Dog B	15.9	20.7	30.2	4.8
	Dog D	14.8	17.1	15.5	2.3
	Rabbit D	46.4	47.1	1.5	0.7
	Rabbit E	14.6	17.4	19.2	2.8
Theophylline					
	Dog C	346	354	2.3	8.0
	Dog D	367	373	1.6	6.0
Procainamide					
	Rabbit F	116	242	109	126
	Rabbit G	81.0	161	98.5	80
	Rabbit H	67.1	133	98.7	65.9

* representing mean transit time of the drug in the sampling tissue

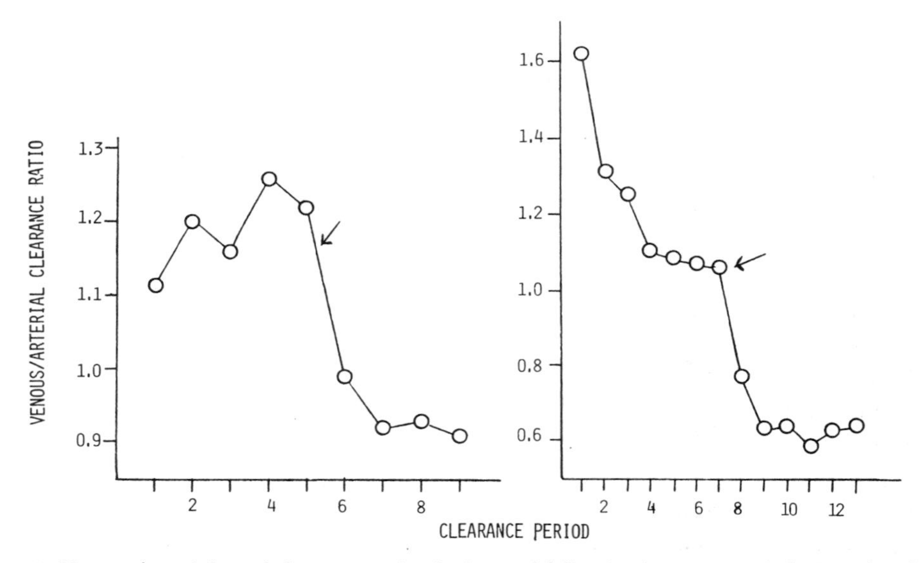

Fig. 11. Venous/arterial renal clearance ratios during and following intravenous infusion of inulin to subject P.L. (left figure; data reproduced from [5]) and of procainamide to a rabbit (right figure; data reproduced from [1]). Arrow signs indicate the end of infusion.

Table II. Comparison of Steady-State Volume of Distribution of Three Drugs Determined Based on Arterial or Venous Plasma Data [2]

Drug	Animal	Arterial V_{ss} (L/Kg)	Venous V_{ss} (L/kg)	Difference (%)
Propranolol				
	Dog	1.5	3.1	107
	Dog	1.8	2.4	33
	Rabbit	5.9	8.6	48
	Rabbit	12.4	18.9	52
Griseofulvin				
	Dog	1.42	1.72	21
	Dog	1.24	1.45	17
	Dog	0.79	0.84	6.3
	Rabbit	2.36	3.61	53
Procainamide				
	Rabbit	1.38	2.80	103
	Rabbit	1.60	3.90	144
	Rabbit	1.20	2.40	100

tration at the common sampling site is always zero, not the highest after an intravenous injection at time zero. The assumption of an instantaneous input may lead to a marked over-estimation of *AUC*; as high as 100% or more over-estimation has been reported [1,2].

5. Contrary to the classical concept, the plasma concentration versus time profile following an intravenous bolus injection actually resembles that obtained after a short-term intravenous infusion. Peak times may occur from 1 to 5 minutes after dosing (Figs. 4, 6, 7). The disposition function of a drug may vary considerably with the sampling site employed (Figs. 7, 8); this is in sharp contrast to the conventional concept of only one disposition function for a given drug in a subject.

6. The size and number of the compartment as well as the micro rate constants in conventional multi-compartmental analysis may vary tremendously with the sampling site chosen. The scientific significance of these micro constants becomes highly questionable.

7. One should also be cautious about the use of the so-called noncompartmental or statistical moment analysis if venous data are employed.

8. Peak time and peak concentration after an extravascular administration may also be sampling-site dependent (Fig. 12). Plasma profiles from different sampling sites may behave like those obtained from entirely different drugs (Figs. 1–8,10).

9. The classical concept of assigning a distribution phase and an elimination phase after a bolus intravenous injection may be extremely misleading if venous data are employed (Figs. 6, 7). The time and concentration at the beginning of the apparent terminal phase may vary greatly with the sampling site. The real distribution phase should be based on arterial

data. The arterial and venous plasma areas under the curve attributed to the terminal exponential phase may differ as much as two times [1,2,18].

10. The exact definition and the calculation method for the initial volume of distribution should be thoroughly reexamined since this term may vary dramatically with the sampling site and also with the use of unrealistically extrapolated zero time concentration or that of the real peak concentration. The extent of initial uptake by the poorly perfused sampling tissue is usually the dominant factor in determining the size of this compartment based on venous data while that by the well-perfused lungs is the dominant factor deciding the size of compartment based on systemic arterial data [1,2].

11. At least one exponential term with a negative coefficient and perhaps also with some lag time is needed to more accurately describe the disposition function of a drug [1,2,32,33].

12. Because of the complexity of dependence of drug concentration on blood sampling site, the validity of using venous plasma data after a single dose to fit into a complex model becomes scientifically questionable.

SIGNIFICANCE IN PHARMACODYNAMICS

The significance of sampling site dependence in drug concentration in pharmacodynamic studies is obvious. First, different plasma concentration profiles obtained from different sampling sites will have different correlations with the same observed pharmacological or toxicological effects in a given study. In other words, the pharmacokinetic/pharmacodynamic correlation may be highly sampling-site dependent. Second, since the "driving force" for drug distribution to receptor sites generally should be the concentration in the arterial blood, strictly speaking, use of changing

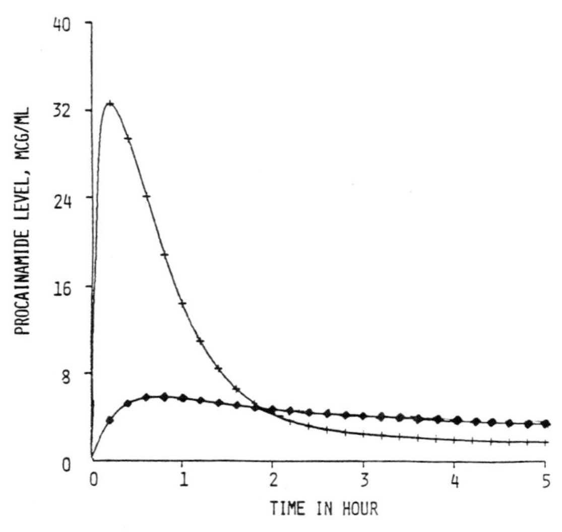

Fig. 12. Simulated arterial (—+—) and venous (—♦—) plasma level-time profiles of procainamide in rabbit #1 [1] following oral absorption (100mg of HCl salt) with a first-order absorption rate constant of 4.16 hour^{-1} (Lam and Chiou, unpublished data).

venous concentration data to correlate with the pharmacodynamic effects would become scientifically questionable in any sophisticated rigorous studies. Many of the reported successful *empirical* corrections between venous data and pharmacodynamic effects may require a reexamination.

For very rapidly acting compounds, the use of venous data may result in a false hypothesis of development of acute tolerance [1,2] as shown in Fig. 13. This is because, compared to arterial levels, venous levels may be relatively low during the absorption or infusion phase and relatively high during the terminal phase (Figs. 2, 6–8). For trichloromonofluoro-methane inhalation in dogs (Fig. 8), at similar venous concentrations during and post inhalation, the arterial levels could differ by almost 10-fold.

The potential marked $A-V$ difference of alcohol (Fig. 3) may have a profound legal implication in the use of alcohol blood level for determining drunkenness in court; this aspect remains to be fully explored.

Use of venous plasma concentration data to predict arterial concentrations shortly after an intravenous injection by assuming a well-stirred sampling compartment has been found to be unsatisfactory in animal studies (unpublished data). This makes the use of venous data for kinetic/dynamic correlation difficult.

CONCLUSIONS

One of the most important data needed in pharmacokinetic and pharmacodynamic studies is drug concentration in blood or plasma. For almost over a century drug concentrations in the blood are commonly assumed to be independent of the sampling site because of the mixing of drugs in the blood circulation is assumed to be extremely rapid or instantaneous. Such an apparent misconception and its implications in pharmacokinetics and pharmacodynamics are briefly reviewed. The importance of the concept of validation of basic assumptions in modeling or in mathematical derivations is emphasized.

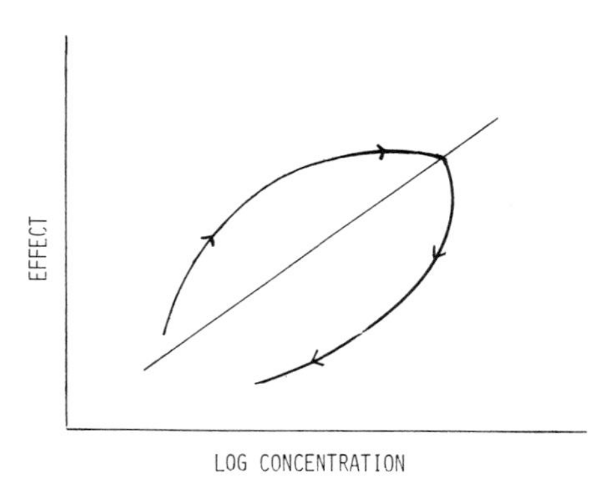

Fig. 13. This figure assumes a true linear relationship between log arterial plasma level and a pharmacological effect. When a marked $A-V$ difference is observed after a short-term infusion or a rapid absorption from an oral dose, use of venous data for correlation with the effect would result in an artifactual clockwise hysteresis loop suggesting development of an acute tolerance (obtained from [2]).

REFERENCES

1. W. L. Chiou. The phenomenon and rationale of marked dependence of drug concentration on blood sampling site. Implications in pharmacokinetics, pharmacodynamics, toxicology and therapeutics. (Part I). *Clin. Pharmacokinet.* **17**:175–199 (1989).
2. W. L. Chiou. The phenomenon and rationale of marked dependence of drug concentration on blood sampling site. Implications in pharmacokinetics, pharmacodynamics, toxicology and therapeutics. (Part II). *Clin. Pharmacokinet.* **17**:275–290 (1989).
3. T. Terada, K. Ishibashi, T. Tsuchiya, H. Noguchi, and T. , Mimura. Arterial-venous concentration gradients as a potential source of error in pharmacokinetic studies. Plasma concentration differences of 6-chloro-2-pyridylmethyl nitrate on constant infusion to rats. *Xenobiotica* **19**:661–667 (1989).
4. W. L. Chiou, G. Lam, M. L. Chen, and M. G. Lee. Arterial-venous plasma concentration differences of six drugs in the dog and rabbit after intravenous administration. *Res. Commun. Chem. Path.* **32**:27–39 (1981).
5. W. L. Chiou and G. Lam. The significance of the arterial-venous plasma concentration difference in clearance studies. *Int. J. Clin. Pharm. Th.* **20**:197–203 (1982).
6. K. B. Bischoff. Physiological pharmacokinetics. *Bull. Math. Biol.* **48**:309–322 (1986).
7. T. K. Henthron, M. J. Avram, and T. C. Krejcie. Intravascular mixing and drug distribution: The concurrent disposition of thiopental and indocyanine green. *Clin. Pharmacol. Ther.* **45**:56–65 (1989).
8. T. Teorell. Kinetics of distribution of substances administered to the body II. The intravascular modes of administration. *Arch. Int. Pharmacod. T.* **57**:226–240 (1937).
9. A. Rescigno and G. Segre. *Drugs and Tracer Kinetics,* Blaisdell Publishing, New York, 1966.
10. S. Riegelman, J. C. K. Loo, and M. Rowland. Shortcomings in pharmacokinetic analysis by conceiving the body to exhibit properties of a single compartment. *J. Pharm. Sci.* **57**:117–123 (1968).
11. M. Gibaldi and D. Perrier. *Pharmacokinetics,* Marcel Dekker, New York, 1982.
12. L. Z. Benet and R. L. Galeazzi. Noncompartmental determination of the steady-state volume of distribution. *J. Pharm. Sci.* **68**:1071–1074 (1979).
13. G. T. Tucker and L. E. Mather. Pharmacokinetics of local anesthetic agents. *Brit. J. Anaesth.* **47**:213–244 (1975).
14. G. T. Tucker and L. E. Mather. Clinical pharmacokinetics of local anesthetics. *Clin. Pharmacokinet.* **4**:241–278 (1979).
15. R. B. Forney, F. W. Hughes, R. N. Harger, and A. B. Richards. Alcohol distribution in the vascular system: Concentration of orally administered alcohol in blood from various points in the vascular system, and in rebreathed air during absorption. *Quart. J. Stud. Alcohol.* **25**:205–220 (1954).
16. S. Bojolm, O. B. Paulson, and H. Flachs. Arterial and venous concentrations of phenobarbital, phenytoin, clonazepam, and diazepam after rapid intravenous injections. *Clin. Pharmacol. Ther.* **32**:478–483 (1982).
17. F. J. Baud, P. Houze, C. Bismuth, A. Jaeger and C. Keyes. Toxicokinetics of paraquate through the heart-lung block. Six cases of acute human poisoning. *J. Toxicol. Clin. Toxi.* **26**:35–50 (1988).
18. G. Lam and W. L. Chiou. Arterial and venous blood sampling in pharmacokinetic studies: Propranolol in rabbits and dogs. *Res. Commun. Chem. Path.* **33**:33–48 (1981).
19. Y. M. Amin, E. B. Thompson, and W. L. Chiou. Fluorocarbon aerosol propellants XII: Correlation of blood levels of trichloromonofluoromethane to cardiovascular and respiratory responses in anesthetized dogs. *J. Pharm. Sci.* **68**:160–163 (1979).
20. W. L. Chiou. A new model-independent physiological approach to study hepatic drug clearance and its applications. *Int. J. Clin. Pharm. Th.* **22**:577–590 (1984).
21. W. L. Chiou. The effect of change in luminal perfusion rate on intestinal drug absorption studied by a simple unified organ clearance approach. *Pharmaceut. Res.* **6**:1056–1059 (1989).
22. W. L. Chiou and H. J. Lee. Effect of change in blood flow on hemodialysis clearance studied by a simple unified organ clearance approach. *Res. Commun. Chem. Path.* **65**:393–396 (1989).
23. A. C. Guyton. *Textbook of Medical Physiology,* 7th ed., W. B. Saunders, Philadelphia, 1986.

24. P. S. Randhawa. Theophylline blood levels in circulatory shock. *Ann. Intern. Med.* **110**:1035 (1989).
25. J. M. Collins and R. L Dedrick. Contribution of lungs to total body clearance: Linear and nonlinear effects. *J. Pharm. Sci.* **71**:66–70 (1982).
26. W. L. Chiou. Potential pitfalls in the conventional pharmacokinetic studies: Effects of the initial mixing of drug in blood and the pulmonary first-pass elimination. *J. Pharmacokin. Biopharm.* **7**:527–536 (1979).
27. M. L. Chen and W. L. Chiou. Pharmacokinetics of methotrexate and 7-hydroxy-methotrexate in rabbits after intravenous administration. *J. Pharmacokin. Biopharm.* **11**:499–513 (1983).
28. H. L. Fung. Pharmacokinetics of nitroglycerin and long-acting nitrate esters. *Am. J. Med.* **74**:13–20 (1983).
29. A. B. Hill, C. J. Bowley, M. L. Nahrwold, P. R. Knight, M. M. Kirsh, and J. K. Denlinger. Intranasal administration of nitroglycerin. *Anesthesiology* **54**:346–348 (1981).
30. W. L. Chiou, G. Lam, M. L. Chen, and M. G. Lee. Effect of arterial-venous plasma concentration differences on the determination of mean residence time of drugs in the body. *Res. Commun. Chem. Path.* **35**:17–26 (1982).
31. R. Haekel. Relationship between intraindividual variation of the saliva/plasma and of the arteriovenous concentration ratio as demonstrated by the administration of caffeine. *J. Clin. Chem. Clin. Bio.* **28**:279–284 (1990).
32. W. L. Chiou, G. Lam, M. L. Chen, and M. G. Lee. Instantaneous input hypothesis in pharmacokinetic studies. *J. Pharm. Sci.* **70**:1037–1039 (1981).
33. M. L. Chen, G. Lam, M. G. Lee, and W. L. Chiou. Arterial and venous blood sampling in pharmacokinetic studies: Griseofulvin. *J. Pharm. Sci.* **71**:1386–1389 (1982).

APPENDIX

A simple mass balance approach to the derivation of Eq. (2).

During the terminal phase, the rate of drug removal from the sampling tissue by venous blood, R_b, should be

$$R_b = (C_v - C_a)QV_s \tag{A–1}$$

where C_v and C_a represent the drug concentration in venous and arterial blood, respectively, Q is the blood flow rate per unit weight of tissue and V_s is the volume of tissue. On the other hand, the rate of disappearance of drug from the sampling tissue, R_d, should be

$$R_d = A_s K \tag{A–2}$$

where A_s is the amount of drug in the sampling tissue and K is the first-order rate constant for the terminal phase. The A_s can also be calculated by

$$A_s = V_s R C_v \tag{A–3}$$

where R is the apparent partition coefficient (more correctly the distribution ratio) of drug between the sampling tissue and the venous blood. Based on the mass balance principle, R_b should be equal to R_d. Combination of the above 3 equations will result in

$$C_a/C_v = 1 - (KR/Q) \tag{A–4}$$

which is identical to Eq. (2).

Derivation of Eq. (3).

Combination of Eqs. (A–1) and (A–2) will yield

$$(C_v - C_a)/C_v = V_s RK/Q \quad \text{(A–5)}$$

If one assumes that V_{area}, the apparent volume of distribution during the terminal phase, also reflects the apparent volume of distribution of the sampling tissue during the terminal phase [1,2], Eq. (A–5) can be written as

$$(C_v - C_a)/C_v = a Varea\, K/Q(A - -6) \quad \text{(A–6)}$$

$$= aCL/Q \quad \text{(A–7)}$$

where a is a proportionality constant and CL is the plasma clearance.

PHARMACODYNAMICS:
MEASUREMENTS AND MODELS

PHYSIOLOGICAL ALTERNATIVES TO THE EFFECT COMPARTMENT MODEL

Nicholas H.G. Holford

Department of Pharmacology
University of Auckland

THE EFFECT COMPARTMENT MODEL

The effect compartment provides a pharmacokinetic model for drug disposition at an effect site. It is not connected in a mass balance sense with the pharmacokinetic model used to describe drug disposition in the whole body.

If the drug concentration in some part of the body, say plasma, is described by a model $C_p(t)$ then the drug concentration in the effect compartment, $C_e(t)$, is given by the solution to the following differential equation.

$$\frac{dC_e(t)}{dt} = K_{1e} \cdot C_p(t) - K_{eo} \cdot C_e(t) \quad (1)$$

The rate in is determined by $C_p(t)$ and a rate constant K_{1e} (from compartment 1 of the model for C_p to the effect compartment) and the rate out is determined by $C_e(t)$ and a rate constant K_{eo} (from the effect compartment to an undefined "outside" compartment).

At steady state the rate in must equal the rate out from the effect concentration compartment. If it is assumed that $C_{p_{ss}} = C_{e_{ss}}$ then:

$$K_{1e} \cdot C_{p_{ss}} = K_{eo} \cdot C_{e_{ss}} \quad (2)$$

so K_{1e} must equal K_{eo}. To emphasize this assumption the symbol K_{eq} is used for both K_{1e} and K_{eo} and Eq. (1) can be rewritten:

$$\frac{dC_e(t)}{dt} = K_{eq}\left[C_p(t) - C_e(t)\right] \quad (3)$$

If $C_p(t)$ is a step-function (e.g., obtained from a loading dose and appropriate maintenance infusion rate), then $C_e(t)$ is:

$$C_e(t) = C_{p_{ss}}\left(1 - e^{-K_{eq}t}\right) \quad (4)$$

C_e will reach 50% of $C_{p_{ss}}$ after one equilibration half-time where:

$$T_{1/2_{eq}} = \ln(2)/K_{eq} \quad (5)$$

Advanced Methods of Pharmacokinetic and Pharmacodynamic Systems Analysis
Edited by D'Argenio, Plenum Press, New York, 1991

THE DISTRIBUTION COMPARTMENT MODEL

The effect compartment model can be interpreted in pharmacokinetic and physiological terms as a drug distribution compartment. It may be understood in a pharmacokinetic sense as a model for drug disposition to a tissue or organ that produces the drug effect. $T_{1/2_{eq}}$ can then be considered as a pharmacokinetic distribution half-life. This is appropriate if the observable drug effect is expected to occur essentially instantaneously once drug molecules reach the site of action.

For example, d-tubocurarine is expected to establish a competitive blockade of post-junctional skeletal muscle acetylcholine receptors quite rapidly (the drug *action*). Observations of muscle twitch (the drug *effect*) may be assumed to reflect the current effects of d-tubocurarine at the neuromuscular junction. If drug effects are delayed with respect to changes in plasma concentration it may be reasonable to assume that the delay is due to the time needed for the drug to diffuse from the plasma to the site of action. The effect compartment model provides a simple description of this phenomenon [1].

The physiological determinants of the distribution half-life can be thought of in terms of the delivery rate to the tissue, i.e., blood flow, and the apparent volume of the tissue, i.e., the actual tissue volume multiplied by the tissue/blood partition coefficient.

Stanski *et al.* [1] provided indirect support for this physiological model by noting that $T_{1/2_{eq}}$ of d-tubocurarine is slower when halothane anesthesia is compared to morphine anesthesia. It is believed that halothane reduces skeletal muscle blood flow in comparison with morphine so it seems reasonable to interpret $T_{1/2_{eq}}$ as a distribution half-life.

A variant of this model is useful in describing the time course of effect in relation to venous concentrations when drug effect is determined by an organ with relatively long arterio-venous transit time. In this case arterial concentrations may be more closely related to the instantaneous drug action. The pharmacokinetic model for the drug in venous blood may be extended to allow drug input to the arterial compartment. The distribution half-time can be used to describe the delayed changes in venous blood concentrations.

THE LAG–TIME MODEL

A delay in onset of sodium excretion has been noted frequently after rapid injection of diuretics. A variant of the pharmacokinetic distribution compartment model may be used to describe this phenomenon. It is known that loop diuretics, such as frusemide, act at the luminal surface of the renal tubule. Drug is delivered to the loop of Henle *via* the urine. Drug may enter the urine both by filtration at the glomerulus and by active secretion in the proximal tubule.

Because it takes a finite time for urine to pass from the glomerulus to the loop of Henle it may be anticipated that there will be a delay in the onset of diuretic effect. A simple model for this delay is provided by a lag time (T_{lag}) between plasma and urine at the site of action. When drug effect is determined from sodium excretion in the urine over a period of time (A_u) a significant error may be introduced in the calculation of the excretion rate if the first collection interval starts at the time of drug administration and a relatively short collection interval (e.g., 20 minutes) is used. If the lag-time for delivery of drug from plasma to the site of action is 5 minutes then the actual excretion interval ($T_1 - T_0$) is closer to 15 minutes because the first five minutes of urine production were determined by events prior to drug administration.

$$\frac{dA_u}{dt} = \frac{A_u}{T_1 - T_0 - T_{lag}} \tag{6}$$

Comparison of an effect compartment model and a lag-time model showed that the predictions of the lag-time model were clearly better because the lag-time not only described the excretion rate of sodium in the first interval but also of frusemide. This provided an explanation for the apparently lower renal clearance of frusemide obtained from the first collection period.

THE PHYSIOLOGICAL SUBSTANCE MODEL

Many drug actions are deduced from observation of changes in physiological substances. In this case the drug effect is monitored by measuring the concentration of the physiological substance. The drug action may be to modify either the synthesis or elimination of the physiological substance.

Synthesis of Physiological Substance

A well studied example is provided by warfarin. Warfarin inhibits vitamin K epoxidase and thus interferes with the re-cycling of vitamin K epoxide to the active vitamin K form. Vitamin K itself is a co-factor in the decarboxylation of precursor proteins for the components of the prothrombin complex of coagulation factors. The half-life of activity of the prothrombin complex has been estimated to be about 14 hours [2]. The time needed for warfarin to distribute to the liver (which is the main site of synthesis of prothrombin complex) and for warfarin to inhibit the epoxidase may be assumed to be very rapid in relation to the half-life of prothrombin complex. In this case the time course of prothrombin complex activity, $PCA(t)$, is given by the solution to:

$$\frac{dPCA(t)}{dt} = R_{syn} \cdot PD\,[W(t)] - K_p \cdot PCA \tag{7}$$

where R_{syn} is the PCA synthesis rate in the absence of warfarin (equivalent to $K_p \cdot PCA_0$, where PCA_0 is the pre-warfarin PCA), K_p is the elimination rate constant for PCA, $W(t)$ is a pharmacokinetic model for warfarin. *PD* is a pharmacodynamic model for the inhibitory effect of warfarin on *PCA* synthesis:

$$PD\,[W] = 1 - \frac{W}{W + IC_{50}} \tag{8}$$

IC_{50} is the warfarin concentration which inhibits synthesis by 50%.

Elimination of Physiological Substance

The time course of plasma potassium changes after administration of a beta-agonist (terbutaline) has been studied by Jonkers *et al.* [3]. The observed delay in changes in potassium with respect to plasma terbutaline concentrations was describable by an effect compartment model. However, the interpretation of the effect compartment model in terms of distribution of terbutaline to an effect site where it is able to produce instantaneous changes in potassium is not reasonable. It is believed that beta-agonists enhance the clearance of potassium from the plasma by stimulating Na-K ATPase at the cell surface. Potassium is then transported into cells and the plasma concentration decreases.

The time course of potassium concentration, $POT(t)$, can be defined with a model similar to that for *PCA* changes caused by warfarin but in this case the drug action is to increase removal rather than decrease synthesis:

$$\frac{dPOT\ (t)}{dt} = R_{in} - POT\ (t) K_{pot} PD\ [Ter(t)] \tag{9}$$

R_{in} is the pre-terbutaline potassium input rate (equivalent to $K_{pot}POT_0$, where POT_0 is the potassium concentration prior to terbutaline administration), K_{pot} is the removal rate constant for potassium (it is assumed that excretion of potassium is negligible over the observed time period) and $Ter(t)$ is a pharmacokinetic model for terbutaline. *PD* is a pharmacodynamic model for the effect of terbutaline on potassium clearance:

$$PD\,[T] = 1 + \frac{E_{max}}{T + EC_{50}} \tag{10}$$

where E_{max} is the maximum stimulation of clearance relative to the pre-terbutaline value and EC_{50} is the terbutaline concentration producing 50% stimulation.

THE INHIBITORY EFFECT COMPARTMENT MODEL

An extension of the effect compartment model has been used by Sheiner [4] to describe tolerance to the effects of nicotine on heart rate. It is proposed that a hypothetical metabolite of nicotine is formed which is an antagonist of nicotine. The time course of the metabolite will therefore modulate the effects of nicotine. While this has provided a descriptive model with interesting predictions there is no evidence for any such antagonist metabolite.

For example, the inhibitory effect of the metabolite, C_m, may be expressed as follows:

$$PD\,[C] = \frac{E_{max}/(1 + C_m(t))}{C + EC_{50}} C \tag{11}$$

$$\frac{dC_m(t)}{dt} = K_m(C(t) - C_m(t)) \tag{12}$$

where C is the active drug concentration and K_m is a formation rate constant for the metabolite from C.

THE PHYSIOLOGICAL FEEDBACK MODEL

A more physiologically based model to account for tolerance can be developed in terms of the regulation of neurotransmitter concentration in the synapse.

Cocaine reduces the removal of transmitters such as noradrenaline and dopamine by inhibiting the neuronal uptake mechanism. Neurotransmitter concentration in the synapse (NT_s) will rise at first with a time course determined by the release rate of transmitter (R_{rel}) and the rate constant for uptake (K_{up}). Feedback systems located on the nerve terminal that respond to increased NT_s by inhibiting R_{rel} are well known. If NT_s increases the synthesis of a release inhibitor (RI) within the nerve terminal, starting from a pre-drug rate of RI_{syn}, then the time course of feedback inhibition which is manifested as tolerance will be determined by a rate constant for the removal of RI from the nerve terminal (K_{tol}).

$$\frac{dNT_s(t)}{dt} = R_{rel}/RI(t) - NT_s(t)K_{up}PD\left[C(t)\right] \tag{13}$$

$$\frac{dRI(t)}{dt} = RI_{syn}NT_s(t) - K_{tol}RI(t) \tag{14}$$

The action of cocaine is to decrease the removal of NT_s but its observed effect is determined by NT_s according to a suitable pharmacodynamic model. Tolerance to the effect of cocaine develops as RI accumulates. It is interesting to note that this model predicts an effect falling below the pre-treatment value when the drug is withdrawn if the half-life of RI is longer than that of the drug. This offers an explanation for "hangover" after stimulant drug administration. This prediction is similar to but not identical to that of model proposed by Ekblad and Licko [5].

REFERENCES

1. D. R. Stanski, J. Ham, R. D. Miller, and L. B. Sheiner. Pharmacokinetics and pharmacodynamics of d-tubocurarine during nitrous oxide-narcotic and halothane anesthesia in man. *Anesthesiology* **51**:235–241 (1979).
2. N. H. G. Holford. Clinical pharmacokinetics and pharmacodynamics of warfarin. *Clin. Pharmacokinet.* **11**:483–504 (1986).
3. R. E. Jonkers, C. J. van Boxtel, R. P. Koopmans, and B. Oosterhuis. A non steady-state agonist interaction model using plasma potassium concentrations to quantify the β_2-selectivity of β-blockers. *J. Pharmacol. Exp. Ther.* **249**:297-302 (1989).
4. L. B. Sheiner. Clinical pharmacology and the choice between theory and empiricism. *Clin. Pharmacol. Ther.* **46**:605–615 (1989).
5. E. B. Ekblad and V. Licko. A model eliciting transient responses. *Am. J. Physiol.* **246**:114–121 (1984).

PHARMACOKINETICS/DYNAMICS OF CORTICOSTEROIDS

William J. Jusko

Department of Pharmaceutics
State University of New York at Buffalo

INTRODUCTION

The corticosteroids are important immunosuppressive and antiinflammatory agents used in treatment of numerous pathophysiologic conditions. Their therapeutic use is complicated by numerous adverse effects when used chronically and there is considerable empiricism in selecting appropriate doses.

The initial step in producing biological responses to corticosteroids is the diffusion of unbound drug from plasma into cells for interaction with cytosolic receptors (Fig. 1). Some types of cells or tissues respond very rapidly while others have a lag phase or slow onset of effect caused by a gene-mediated mechanism of action. Both types of responses last considerably beyond the time-course of active drug in the system. For example, prednisolone, a moderately lipid soluble compound, rapidly distributes into various cells and tissues, and has a pharmacokinetic $t_{\frac{1}{2}}$ of about 3 hours and a duration of biologic effects of 18 to 36 hours, depending on dose. Several pharmacokinetic/dynamic models have been proposed to rationalize and quantitate these response patterns.

DIRECT EFFECTS

Relatively simple equations can be used to describe the "direct" or non-gene-mediated effects of corticosteroids on cell trafficking patterns of basophils, T-helper cells, and others (Fig. 2). It is assumed that corticosteroids cause an immediate change in the affinity of cells for sites in an extravascular compartment. The decline in cell number is attributed to inhibition of cell movement from extravascular sites and the blood cell replenishment occurs when corticosteroid concentrations in plasma fall below the IC_{50} value. The IC_{50} value for methylprednisolone effects in man is similar in magnitude to its K_D or drug-receptor equilibrium dissociation constant.

A more recent model for basophil cell trafficking has allowed for more accurate quantitation of cell movement between blood and extravascular sites and permits extrapolation of effects to a wider range of steroid doses [3].

Similar equations and patterns to those shown in Fig. 2 are applicable to adrenal suppressive effects of methylprednisolone with an added complication of the necessity of accounting for the episodic-circadian synthesis and secretion of cortisol which governs the baseline conditions of the endogenous steroid [4]. Figure 3 depicts a pharmacodynamic model for the synthesis and secretion of cortisol (R_{cort}), for which the following equation applies:

Advanced Methods of Pharmacokinetic and Pharmacodynamic Systems Analysis
Edited by D'Argenio, Plenum Press, New York, 1991

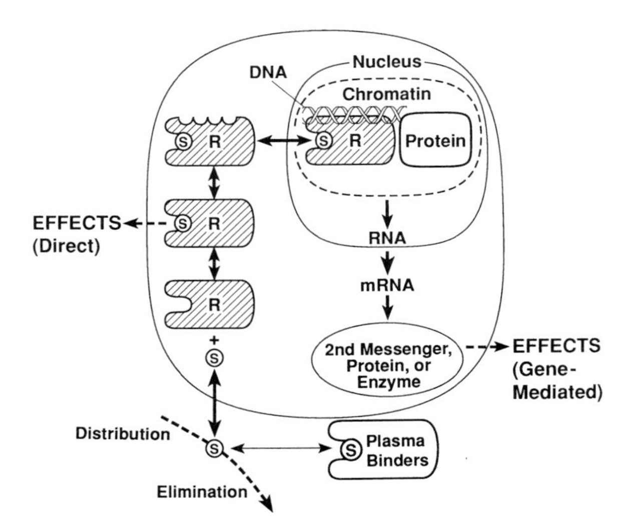

Fig. 1. Receptor (R)-mediated "Direct" and "Gene-Mediated" mechanisms of corticosteroid (S) hormone effects in steroid sensitive cells. From [1].

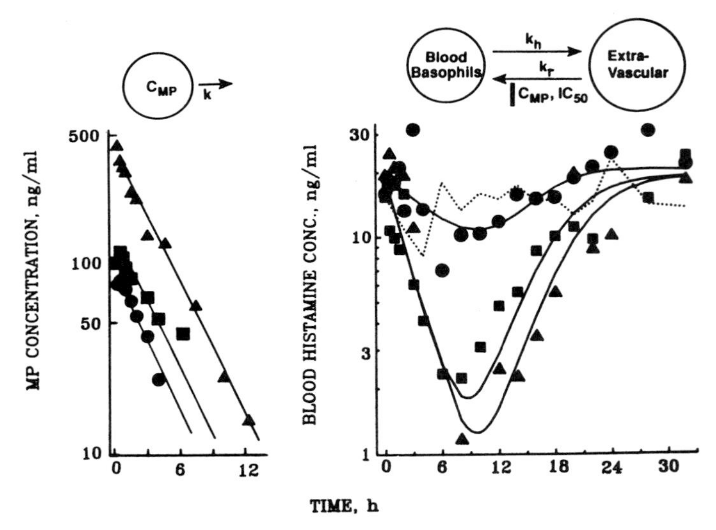

Fig. 2. Pharmacokinetic (left) and two-compartment cell trafficking and direct suppression model for basophils (whole blood histamine) (right) in relation to methylprednisolone (C_{MP}) concentrations where IC_{50} is the C_{MP} producing 50% inhibition of basophil return to blood (k_r). Methylprednisolone concentrations and blood basophils as whole blood histamine concentrations (C_B) versus time were obtained in a normal male subject at baseline (C_B^0, dotted line), and after 10 (circles), 20 (squares) and 40 (triangles) mg of intravenous methylprednisolone (as Solu-Medrol, Upjohn). Solid lines are least-squares simultaneous fitting of data from all dose levels to the relationships:

$$C_{MP} = C_i^0 e^{-kt} \quad \text{and} \quad C_B = C_B^0 e^{-k_h t} + C_B^0 \left(1 - \frac{C_{MP}}{C_{MP} + IC_{50}}\right)$$

where C_i^0 and k are the intercepts and slope of the pharmacokinetic profiles. From [2].

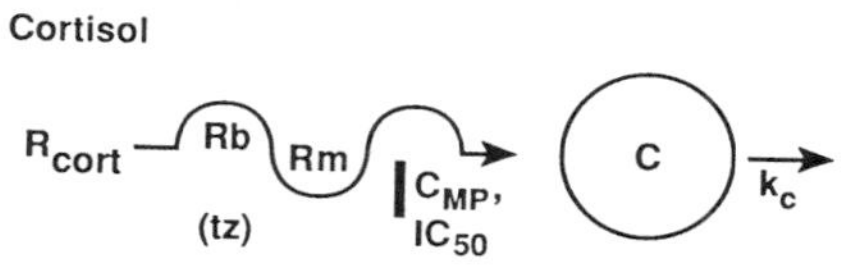

Fig. 3. Pharmacodynamic model for cortisol synthesis and secretion (R_{cort}) governed by a circadian rhythm.

$$R_{cort} = R_m + R_b cos(t - t_z)\frac{15}{57.3} \quad (1)$$

Cortisols inhibition by methylprednisolone (C_{MP}) is given by:

$$C = C_0 e^{-k_c t} + R_{cort}(1 - \frac{C_{MP}}{IC_{50} + C_{MP}}) \quad (2)$$

where C = cortisol concentration, C_0 = initial cortisol concentration, k_c = rate constant for cortisol elimination, and IC_{50} = concentration of methylprednisolone causing 50% inhibition of cortisol influx. In considering optimal dosage regimens, the hypothesis was made that a "steroid-sparing" effect could be achieved by designing dosage regimens such that a loading dose occupies all of the receptors and a maintenance dose refills receptors as they recycle following drug elimination. This was tested successfully in human studies (Fig. 4).

The kinetics and dynamics of methylprednisolone were recently examined in obese and normal men [6]. While obesity produced a slower clearance of the steroid, our pharmacodynamic models allowed determination that the IC_{50} values for steroid effects on cortisol secretion, basophil trafficking, and T-helper cell movement were unaltered in overweight men.

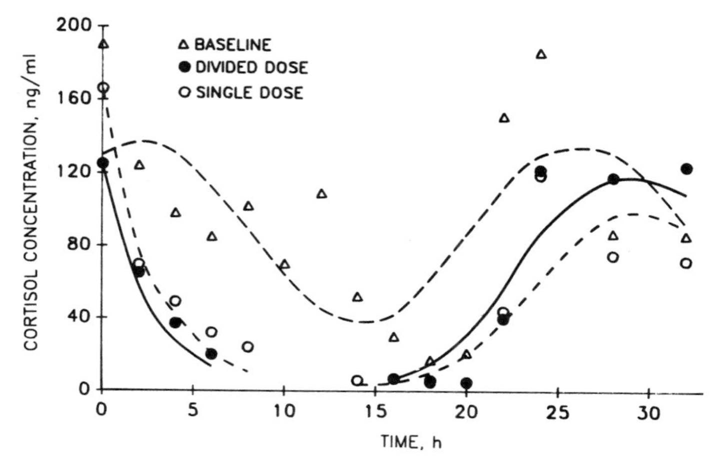

Fig. 4. Plasma cortisol concentrations versus time for one human subject during a baseline period and following administration of a single 40 mg bolus dose of methylprednisolone and a divided dose of 20 mg bolus and 5 mg 8 hours later. Symbols are experimental data and lines are fitted model predictions (Eqs. (1) and (2)). From [5].

GENE-MEDIATED EFFECTS

First- [7] and second-generation [8] models were developed to account for the gene-mediated effects of prednisolone in rats. According to the model in Fig. 1, after access of drug to cytosolic receptors, the drug-receptor complex is activated and rapidly translocated into the nucleus where it binds to glucocorticoid response elements (GRE) on DNA. These regulate and usually increase the transcription of specific genes thereby modulating mRNA concentrations and affecting synthesis of various specific proteins, second messengers, or enzymes. For example, the tyrosine aminotransferase enzyme (*TAT*) is a commonly studied marker of such effects. In our experiments to model gene-mediated corticosteroid effects, *in vivo* measurements included the steroid in plasma, free hepatic cytosolic receptors, and the tyrosine aminotransferase (*TAT*) enzyme in liver after iv doses of prednisolone. Separate *in vitro* quantitation of prednisolone-receptor k_{on} and k_{off} values was also done. The following differential equations (from [8]) describe drug in plasma, free receptors, bound receptors, receptor binding to DNA, transit compartments symbolic of RNA and mRNA, and *TAT* activity as a function of time (see Fig. 5):

$$D = C = \sum C_i e^{-\lambda_i t} \tag{3}$$

$$\frac{d(R)}{dt} = -k_{on}(D)(R) = k_{off}(DR) + DRN \tag{4}$$

$$\frac{d(DR)}{dt} = k_{on}(D)(R) - (k_{off} + k_{DNA})(DR) \tag{5}$$

$$\frac{d(DRN)}{dt} = [k_{DNA}(DR) - DRN]/\tau \tag{6}$$

$$\frac{d(M_1)}{dt} = [DRN - M_1]/\tau \tag{7}$$

$$\frac{d(M_2)}{dt} = [M_1 - M_2]/\tau \tag{8}$$

$$\frac{d(TAT)}{dt} = E_f M_2^p - k_{de}(TAT) \tag{9}$$

where D and R are drug concentrations in plasma and at hepatic cytosol receptor sites, respectively; C_i and λ_i are intercept and slope coefficients; *DR* is drug-receptor complex concentration; *DRN* is drug-receptor-DNA complex concentration; M_1 and M_2 are intermediary transfer compartments, *TAT* is the hepatic tyrosine aminotransferase activity; k_{on} and k_{off} are the rate constants for drug receptor association and

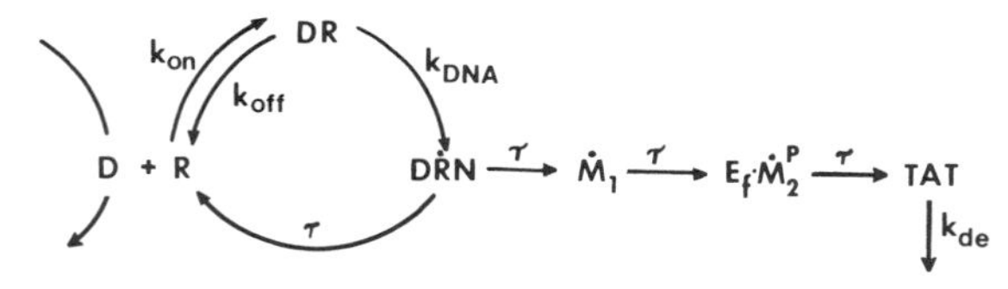

Fig. 5. Pharmacodynamic model and equations for gene-mediated glucocorticoid action. (See text for symbol definitions.)

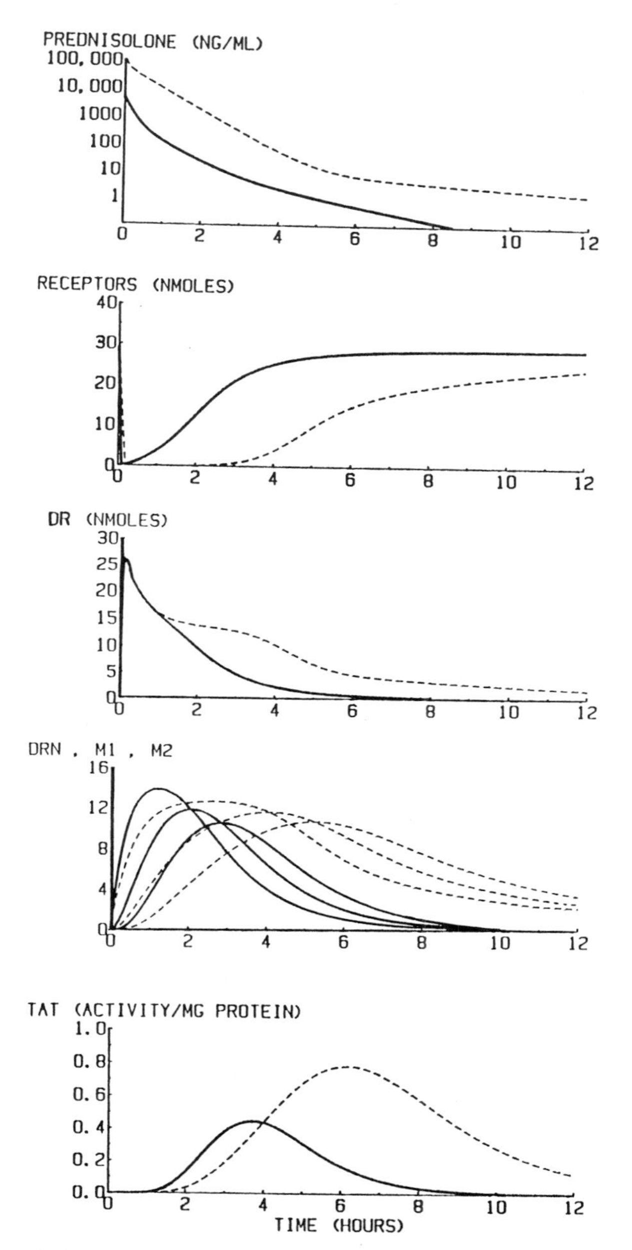

Fig. 6. Time-course of plasma concentrations of prednisolone (top), free and bound (DR) receptor concentrations, transfer compartments (DRN, M_1, M_2) and hepatic *TAT* activity (bottom) following iv administration of 5 (solid) and 50 (dash line) mg/kg prednisolone in rats. The lines represent least-squares fitting of data by the model in Fig. 5. From [8].

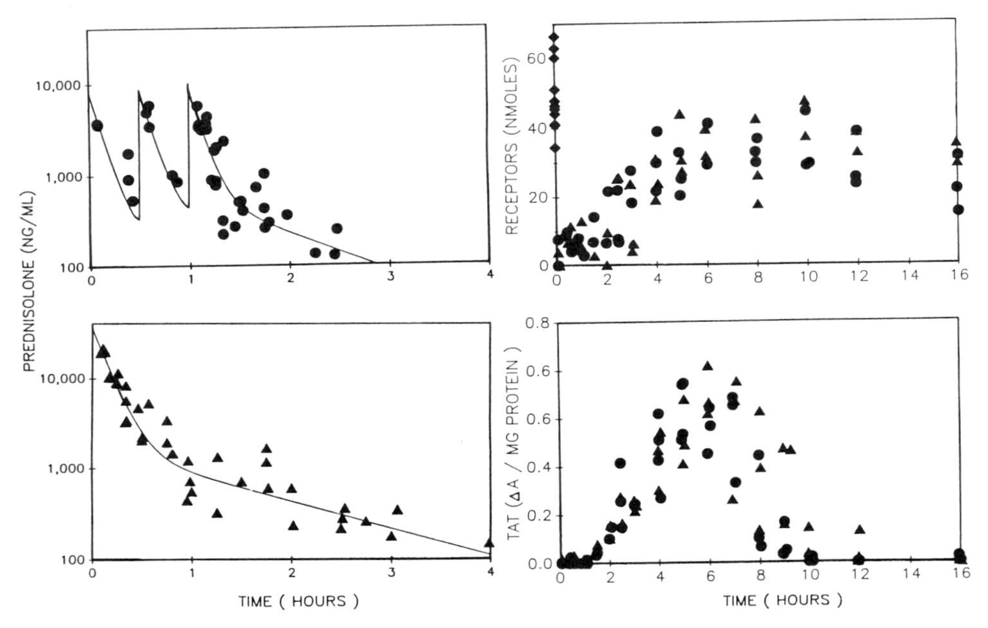

Fig. 7. Results of a *steroid-sparing* study of prednisolone regimens in rats. Pharmacokinetics of a single 25 mg/kg bolus dose (triangles) versus three 5 mg/kg doses (circles) are shown (left) along with hepatic receptor and *TAT* profiles (right). From [10].

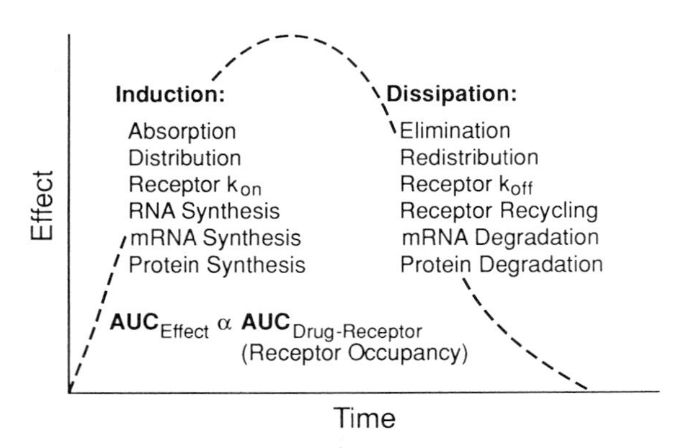

Fig. 8. Glucocorticoid Pharmacodynamics. Summary of major determinants of the time-course of onset and dissipation of corticosteroid effects and the overall *AUC* of Effect. From [1].

dissociation; k_{DNA} is the rate constant for DR binding to DNA; λ is a transfer time parameter; E_f is an efficiency factor; P is a power term; and k_{de} is a rate constant for degradation of TAT. This model was applied in one study to data obtained after 5 and 50 mg/kg doses of prednisolone in adrenalectomized rats [7,8] to obtain the profiles shown in Fig. 6.

Another study assessed the effect of obesity on prednisolone disposition and dynamics in rats [9]. A third study compared the responses to a single 25 mg/kg bolus dose versus three 5 mg/kg doses of prednisolone given at 30 min. intervals [10]. A "steroid sparing" effect of equal receptor occupancy and TAT response was achieved with the second dosage regimen (Fig. 7). This provides promise that pharmacodynamic models will allow improved rationalization of steroid dosage regimens.

Recent studies have similarly evaluated methylprednisolone kinetics and dynamics in rats [11]. This steroid is more potent than prednisolone and caused a marked degree of receptor down-regulation which was successfully modeled by extending our previous model and equations (Fig. 5 and Eqs. (3)–(9)).

SUMMARY

Major conclusions from these studies are that receptor occupancy directly governs biologic responses to corticosteroids (Fig. 8) and that the slow onset of effects is caused not by pharmacokinetic factors, but by either cell redistribution with an elapse of time required to achieve maximum suppression of the direct effects (Fig. 2) or the time needed for mRNA, protein, enzyme, or second messenger synthesis (Fig. 6). The slow dissipation of steroid effects is governed by steroid elimination (time to decline below IC_{50} or K_D values), the k_{off} of receptor binding, and the time required for biologic factors or cells to readjust to baseline or equilibrium conditions. These studies have allowed considerable insight to be gained concerning the integrated role of drug, receptors, and biologic response mediators in determining the role of dose, time, the type of corticosteroid, and drug/disease interactions in governing the pharmacodynamics of these agents.

ACKNOWLEDGMENTS

This work was supported by NIH Grant GM 24211. Presentation of this material at the BMSR Workshop on *Advanced Methods of Pharmacokinetic and Pharmacodynamic Systems Analysis* on May 18–19, 1990 by Mr. Jeffrey A. Wald is appreciated.

REFERENCES

1. W. J. Jusko. Corticosteroid pharmacodynamics: Models for a broad array of receptor-mediated pharmacologic effects. *J. Clin. Pharmacol.* **30**:303–310 (1990).
2. A-N. Kong, E. A. Ludwig, R. L. Slaughter, P. M. DiStefano, J. DeMasi, E. Middleton Jr. , and W. J. Jusko. Pharmacokinetics and pharmacodynamic modeling of direct suppression effects of methylprednisolone on serum cortisol and blood histamine in human subjects. *Clin. Pharmacol. Ther.* **46**:616–628 (1989).
3. J. A. Wald and W. J. Jusko. Two-compartment basophil cell trafficking model for methylprednisolone pharmacodynamics. *Pharmaceut. Res.* **7**:S-219 (1990).
4. W. J. Jusko, W. R. Slaunwhite, and T. Aceto Jr.. Partial pharmacodynamic model for the circadian-episodic secretion of cortisol in man. *J. Clin. Endocrinol. Metab.* **40**:278–289 (1975).

5. W. G. Reiss, R. L. Slaughter, E. A. Ludwig, E. Middleton, and W. J. Jusko. Steroid dose sparing: Pharmacodynamic responses to single versus divided doses of methylprednisolone in man. *J. Allergy Clin. Immun.* **85**:1058–1066 (1990).
6. T. Dunn, E. Ludwig, R. Slaughter, D. Camara, and W. J. Jusko. Comparative pharmacokinetics and pharmacodynamics of methylprednisolone in obese and non-obese men. *Clin. Pharmacol. Ther.* **47**:181 (1990).
7. F. D. Boudinot, R. D'Ambrosio, and W. J. Jusko. Receptor-mediated pharmacodynamics of prednisolone in the rat. *J. Pharmacokin. Biopharm.* **14**:469–493 (1986).
8. A. I. Nichols, F. D. Boudinot, and W. J. Jusko. Second generation model for prednisolone pharmacodynamics in the rat. *J. Pharmacokin. Biopharm.* **17**:209–227 (1989).
9. A. I. Nichols, R. D'Ambrosio, and W. J. Jusko. Pharmacokinetics and pharmacodynamics of prednisolone in obese rats. *J. Pharmacol. Exp. Ther.* **250**:963–970 (1989).
10. A. I. Nichols and W. J. Jusko. Receptor mediated prednisolone pharmacodynamics in rats: Model verification using a dose-sparing regimen. *J. Pharmacokin. Biopharm.* **18**:189–208 (1990).
11. D. B. Haughey and W. J. Jusko. Receptor-mediated pharmacodynamics of methylprednisolone in the rat. *Pharmaceut. Res.* **7**:S-240 (1990).

VARIABILITY IN HUMAN CARDIOVASCULAR PHARMACODYNAMICS

Darrell R. Abernethy

Program in Clinical Pharmacology
Brown University

INTRODUCTION

An often-stated comment from researchers in aging is that the consistent hallmark of increasing age is increasing variability for measured parameters within a population. The various components of cardiovascular function in the aging human generally fit this maxim. Here an attempt will be made to characterize aspects of pharmacodynamic variability in aging and explore some of the various components of physiological variability that contribute to measurable pharmacodynamic endpoints. The approach will stress that useful cardiovascular pharmacodynamic endpoints (e.g., blood pressure, heart rate, cardiac output, etc.) are complex parameters derived from interacting components. Blood pressure, for example, characterized in terms of hemodynamics is a function of cardiac output, vascular compliance, and vascular resistance. Cardiac output is comprised of heart rate and contractility, while contractility may be usefully separated into systolic and diastolic components.

Approaches to characterize observed variance for such composite parameters require exploring the relative contributions to the overall variance from the components that are amenable to measurement. However, developing estimates of overall variance for the composite parameter (e.g., blood pressure) from measured component parameters is not straightforward, as these contributions to overall variance may not be simply additive or multiplicative, and in addition may vary with physiological state. For example heart rate variability associated with respiration and vagal tone at rest (predominantly a function of parasympathetic tone) is often quite predictable and marked. However, it may disappear completely during exercise when heart rate is increased and much more regulated by an activated sympathetic nervous system. With this background this paper will address the following issues.

1. Age-related physiological parameters that may predict *in vivo* variability in cardiovascular pharmacodynamics.

2. Examples of changes in basal state or "start point" that impact on variability in cardiovascular pharmacodynamics.

3. Examples of time-related state changes that impact on cardiovascular pharmacodynamic variability.

AGE-RELATED CHANGES IN CARDIOVASCULAR PHYSIOLOGY THAT ALTER PHARMACODYNAMICS

Cardiac

Resting and exercise cardiac output may not change with age in carefully selected healthy elderly individuals [1], however, one may assume an age-related decline in a usual clinical population and a more marked decline in aged patients with hypertension and/or coronary artery disease [2]. Components of the cardiac cycle change with aging, such that heart rate decreases, end diastolic volume increases, and the importance of late diastolic filling of the left ventricle increases. The central cardiac physiologic alteration in the aged individual is impairment of diastolic relaxation which results in the observed changes in cardiac cycle [3]. An important consequence of these age-related changes is an impaired capacity to tolerate high heart rate and increased importance of left atrial contraction to augment left ventricular filling. Pharmacological maneuvers which result in marked increase in heart rate such as sympathetic nervous stimulation or parasympathetic blockade, then can be predicted to have a different pharmacodynamic outcome for a given change in heart rate when comparing elderly to young individuals. Elderly individuals may have less increase in cardiac output due to limitation in left ventricular filling resulting from the decreased diastolic filling time.

Peripheral Vascular

Hallmarks of vascular aging are decreased vascular compliance and increased peripheral vascular resistance [4,5]. Changes in compliance have been attributed to arterial structural changes with increased cellular matrix and fibrous tissue leading to decreased vascular distensibility. Increased peripheral vascular resistance has been repeatedly demonstrated in aging hypertensive and normotensive individuals, however, the mechanism responsible for this finding is not well understood. The relevant result of decreased vascular compliance in the elderly is increased systolic blood pressure, which will provide a basis for discussion of the comparison of a measured pharmacodynamic parameter when the initial starting point is different, here systolic blood pressure in elderly versus young individuals.

Adrenergic Nervous System

Perhaps the most extensively characterized age-related change in the cardiovascular system is impairment of beta-adrenergic responsiveness, both cardiac (β_1) and peripheral vascular (β_2) [6,7]. Mechanism for this observation has been extensively studied and remains incompletely understood, however, the functional impairment is post-receptor within the β-adrenergic signal transduction cascade [8]. The *in vivo* result is marked blunting of β_1-adrenergic cardiac chronotropic stimulation and β_2-adrenergic peripheral vasorelaxation in aged individuals. Pharmacodynamic consequences are decreased responsiveness to both β-adrenergic stimulation and inhibition in aged individuals. In contrast, extensive *in vivo* and *in vitro* study indicates alpha-adrenergic function is unchanged in aging [9,10].

Renin-Angiotensin-Aldosterone Axis

Aging is associated with decreased plasma renin activity, leading some to postulate that this axis has a limited role in cardiovascular homeostasis in the elderly. However, demonstration of a local vascular renin-angiotensin axis that may have increased function in aging has complicated interpretation of the age-related role of this system [11]. Pharmacodynamic studies have been useful to explore this question *in vivo.* Blockade of the renin-angiotensin system with converting enzyme inhibitors has similar hypotensive effect in aged as compared to younger individuals, suggesting a little change in the role of angiotensin II in the maintenance of peripheral vascular tone [12]. However, decreased concentrations and responsiveness to stimulation of aldosterone in the elderly may provide a basis for the lack of fluid retention due to secondary hyperaldosteronism when vasodilators are administered to elderly patients [13].

Atrial Natriuretic Factor, Endothelial Derived Relaxing Factor, Endothelin

An ever-increasing number of endogenous substances, both circulating (endocrine) and locally active (paracrine) which affect cardiovascular function when administered systemically to humans are being described [14–16]. Most have either vasorelaxing (atrial natriuretic factor, endothelial derived relaxing factor) or vasoconstricting (endothelin) effects and some may alter renal sodium and water excretion (atrial natriuretic factor) when administered pharmacologically. The importance of these factors in maintenance of cardiovascular function is incompletely defined at this time, and will only become clear as pharmacological inhibitors of their function are developed. The age-associated changes in these cardiovascular-active substances are presently under investigation, however, at the present time no conclusions can be drawn, either from the physiological observations or their pharmacodynamic consequences.

CHANGE IN BASAL STATE THAT IMPACTS ON VARIABILITY IN CARDIOVASCULAR PHARMACODYNAMICS

Systolic blood pressure will be used as an example. Aged hypertensive and normotensive individuals consistently have higher and more variable systolic blood pressure than younger individuals when groups are selected for similar diastolic

Table I. Comparison of Characteristics of Young and Elderly Hypertensive Patients

Patient characteristics	Young (n=12)	Elderly (n=12)
Age (yr)	35±1	71±1*
Weight (kg)	90±3	86±5
Sex (M,F)	10M,2F	10M,2F
Race (W,B)	8W,4B	12W, 0B
Baseline systolic blood pressure (mm Hg)	146±3	175±5*
Baseline diastolic blood pressure (mm Hg)	103±2	99±1
Baseline heart rate (beats/min)	69±2	70±3
Baseline PR interval (msec)	159±5	186±11*

Data are mean values ±SE; W, white; B, black. * $p < 0.05$ by Student T test. From [18] (reprinted with permission).

blood pressure (Table I) [17]. Interpretation and comparison of pharmacodynamics of antihypertensive drugs across age may become complex, as measurement and comparison of absolute effect, which may be of greatest pharmacological interest, will be derived from non-comparable populations, both in measured parameter and its variance. A number of transformations, including normalization to percent change, and evaluation of mean blood pressure

$$\frac{SBP - DBP}{3} + DBP \quad (1)$$

where SBP=systolic blood pressure and DBP=diastolic blood pressure have been proposed, none with totally satisfactory maintenance of the measured differences in variance noted in experimental observation. Attempts to stabilize variance by increasing numbers of measures and therefore information density (e.g., beat-to-beat blood pressure and 24 hour blood pressure monitoring) have simply demonstrated in more conclusive fashion that aged versus young populations are truly not with comparable variance and this is not due to sampling error. Interpretation of drug effects on systolic blood pressure and comparison between aged and young individuals remains a conceptual challenge.

TIME-RELATED STATE CHANGES THAT IMPACT ON CARDIOVASCULAR PHARMACODYNAMIC VARIABILITY

Effect of a calcium antagonist drug on decrease in mean blood pressure in young and aged hypertensive patients after acute intravenous drug exposure, short term (2 week) oral drug administration, and long term (12 week) oral drug administration will be used as an example [19]. Using a linear pharmacokinetic-pharmacodynamic model the aged patients have greater hypotensive effect to a given drug concentration than do younger patients (Fig. 1.). After 2 weeks of exposure to the same drug this relationship has changed and the pharmacodynamic effect is similar in young and elderly. The slope of the relationship indicates the young now have increased effect that approximates that in the aged after the initial dose, and that the aged patients have had no change in pharmacodynamic effect (Fig. 2.). With 3 month drug exposure, the kinetic-dynamic relationship is similar for young and aged, however, the slope of the relationship indicates both groups are less "sensitive" to a given drug concentration than was observed at the 2 week interval (Fig. 3.). With this "state" change over time, a simple pharmacodynamic model effectively characterizes the data, however, interpretation of the fitted data is markedly different depending on the chosen time window observation. Mechanism for these observations is unclear and a focus of ongoing investigation.

This is but one example from our laboratory of an apparent "state" change over time that is different in young versus aged individuals, another prominent example may be β-adrenergic regulation and desensitization as a function of time of exposure to agonist stimulation [20]. Finally, an elegant series of animal studies has demonstrated disease-related changes in pharmacokinetic-pharmacodynamic relationships, raising the likelihood that "state" changes over time as a function of onset or effective treatment of underlying disease occur, and require development of pharmacodynamic models to accommodate such time-related change [21].

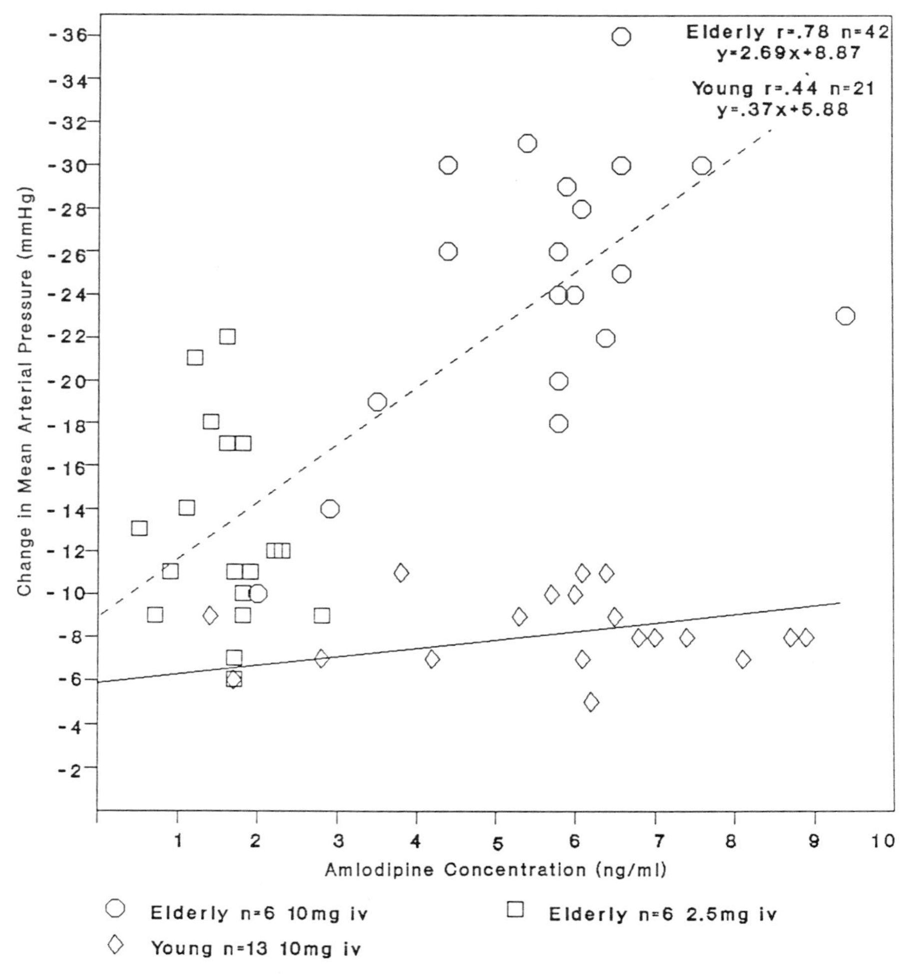

Fig. 1. Amlodipine intravenous pharmacodynamics 0.5–96 hrs following the 1st dose. Mean amlodipine concentration at each sampling time point (abscissa) in relationship to decrease in mean arterial pressure from pretreatment baseline from 0.5 to 96 hours after the initial intravenous dose. Lines represent the least-squares linear regression analysis. Octagons—elderly, n=6, dose 10mg intravenously, diamonds—young, n=13, 10mg intravenously, squares—elderly, n=6, 2.5mg intravenously. From [19].

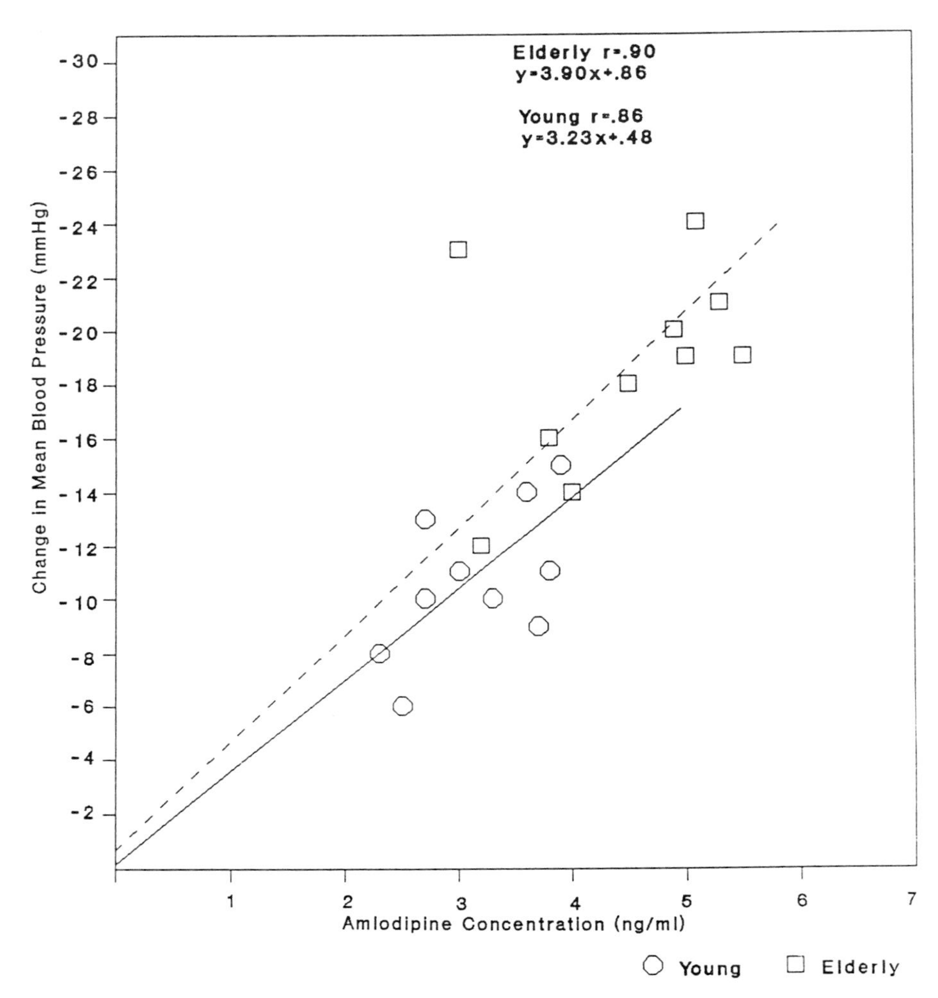

Fig. 2. Amlodipine 2-week pharmacodynamics 0–24 hrs following 2.5mg dose. Mean amlodipine concentration at each sampling time point (abscissa) in relationship to decrease in mean arterial pressure from pretreatment baseline after 2 weeks amlodipine treatment at a dose of 2.5mg p.o. daily.

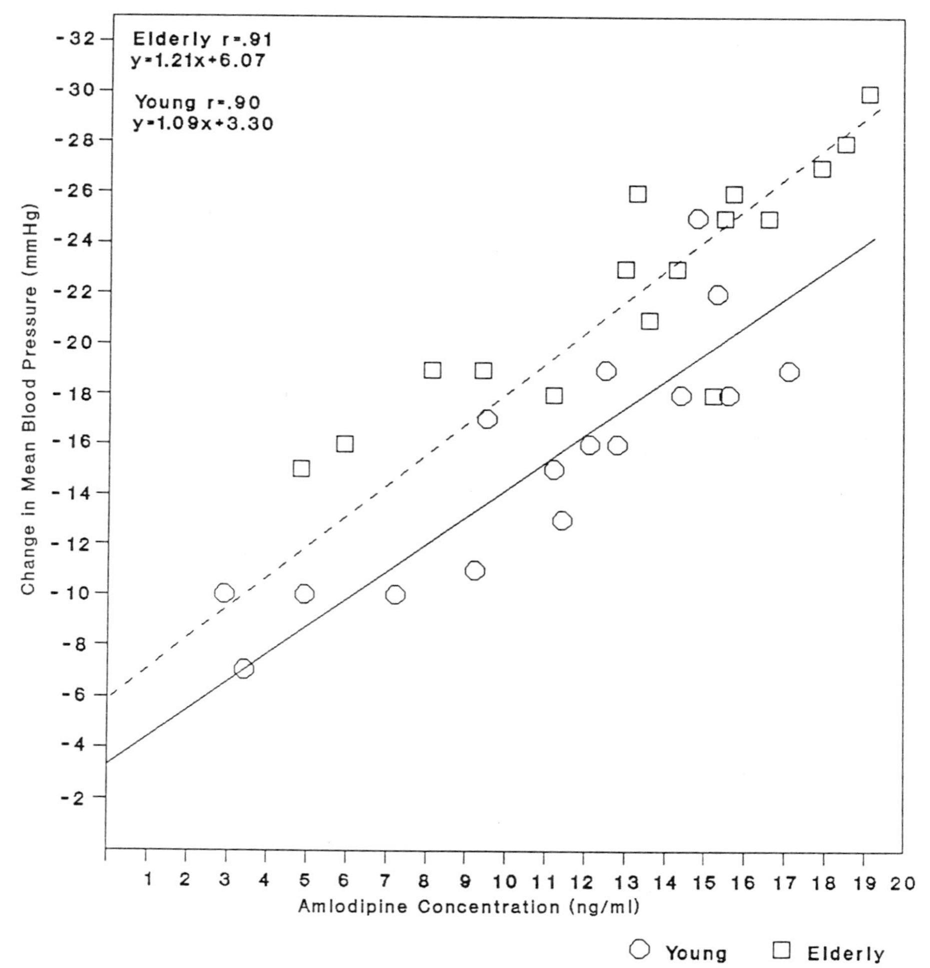

Fig. 3. Amlodipine 14-week pharmacodynamics 0–144 hrs following last dose with all dose levels combined. Mean amlodipine concentration at each time point (abscissa) in relationship to decrease in mean arterial pressure from pretreatment baseline after long term amlodipine treatment. Octagons and solid line represent young patients and squares and broken line elderly patients.

CONCLUSION

Aging humans have provided a useful example and source of experimental data to explore and define sources of pharmacodynamic variance in a reasonably defined system. Cardiovascular pharmacodynamics have provided a particularly useful system for study due to the extensive description of age-and disease-related changes in cardiovascular physiology that may contribute to pharmacodynamic variance. However, experimental and clinical situations such as different "start points" and time related "state" changes in a measurable pharmacodynamic parameter are incompletely accommodated using present pharmacokinetic-pharmacodynamic models. This offers an intriguing and useful set of challenges for future model development.

ACKNOWLEDGMENT

Supported in part by DHHS Grant No. AG-08206. The author gratefully acknowledges Grazina Kulawas for administrative support and manuscript preparation.

REFERENCES

1. R. J. Rodeheffer, G. Gerstenblith, L. C. Becker, J. L. Fleg, M. L. Weisfeldt, and E. G. Lakatta. Exercise cardiac output is maintained with advancing age in healthy human subjects: cardiac dilatation and increased stroke volume compensate for diminished heart rate. *Circulation* **69**:203–213 (1984).
2. R. O. Bonow, S. L. Bacharach, M. V. Green, K. M. Kent, D. R. Rosing, L. C. Lipson, M. B. Leon, and S. E. Epstein. Impaired left ventricular diastolic filling in patients with coronary artery disease: assessment with radionuclide angiography. *Circulation* **64**:315–323 (1981).
3. R. O. Bonow, D. F. Vitale, S. L. Bacharach, B. J. Maron, and M. V. Green. Effects of aging on asynchronous left ventricular regional function and global ventricular filling in normal human subjects. *J. Am. Coll. Cardiol.* **11**:50–58 (1988).
4. A. P. Avolio, S. G. Chen, R. P. Wang, *et al.* Effects of aging on changing arterial compliance and left ventricular load in a northern Chinese urban community. *Circulation* **68**:50–58 (1983).
5. F. H. Messerli, K. Sundgaard-Riise, H. O. Ventura, F. G. Dunn, L. B. Glade, and E. D. Frolich. Essential hypertension in the elderly: hemodynamics, intravascular volume, plasma renin activity, and circulating catecholamine levels. *Lancet* **2**:983–986 (1983).
6. R. E. Vestal, A. J. J. Wood, and D. G. Shand. Reduced beta-adrenoceptor sensitivity in the elderly. *Clin. Pharmacol. Ther.* **26**:181–186 (1979).
7. H. Y. Pan, B. B. Hoffman, R. A. Pershe, and T. F. Blaschke. Decline in beta-adrenergic receptor-mediated vascular relaxation with aging in man. *J. Pharmacol. Exp. Ther.* **239**:802–807 (1986).
8. J. H. Chin and B. B. Hoffman. Age-related deficit in beta receptor stimulation of cAMP binding in blood vessels. *Mech. Ageing Dev.* **53**:111–125 (1990).
9. C. Klein, J. G. Gerber, N. A. Payne, and A. S. Nies. The effect of age on the sensitivity of the α-1-adrenoceptor to phenylephrine and prazosin. *Clin. Pharmacol. Ther.* **47**:535–539 (1990).
10. P. J. W. Scott and J. C. Reid. The effect of age on the responses of human isolated arteries to noradrenaline. *Brit. J. Clin. Pharmaco.* **13**:237–239 (1982).
11. V. J. Dzau. Multiple pathways of angiotensin production in the blood vessel wall: evidence, possibilities, and hypotheses. *J. Hypertens.* **7**:933–936 (1989).
12. G. H. Williams. Converting-enzyme inhibitors in the treatment of hypertension. *N. Engl. J. Med.* **319**:1517–1525 (1988).
13. R. H. Noth, M. N. Lassman, and S. V. Tan. Age and the renin-aldosterone system. *Arch. Intern. Med.* **137**:1414–1420 (1977).
14. D. J. Webb, N. Benjamin, M. J. Allen, J. Brown, M. O'Flynn, and J. R. Cockroft. Vascular responses to local atrial natriuretic peptide infusion in man. *Brit. J. Clin. Pharmaco.* **26**:245–251 (1988).

15. P. Vallance, J. Collier, and S. Moncada. Effects of endothelium-derived nitric oxide on peripheral arteriolar tone in man. *Lancet* **2**:997–1000 (1989).

16. J. G. Clarke, N. Benjamin, S. W. Larkin, D. J. Webb, G. J. Davies, and A. Maseri. Endothelin is a potent long-lasting vasoconstrictor in men. *Am. J. Physiol.* **257**:H2033–H2035 (1989).

17. C. Garland, E. Barrett-Connor, L. Suarez, and M. H. Criqui. Isolated systolic hypertension and mortality after age 60 years, A prospective population-based study. *Am. J. Epidemiol.* **118**:365–376 (1983).

18. S. C. Montamat and D. R. Abernethy. Calcium antagonists in geriatric patients: Diltiazem in elderly hypertensives. *Clin. Pharmacol. Ther.* **45**:682–691 (1989).

19. D. R. Abernethy, J. Gutkowska, and L. M. Winterbottom. Effects of amlodipine, a long-acting dihydropyridine calcium antagonist in aging hypertension: Pharmacodynamics in relation to disposition. *Clin. Pharmacol. Ther.* **48**:76–86 (1990).

20. J. Fraser, J. Nadeau, D. Robertson, and A. J. J. Wood. Regulation of human leukocyte beta receptors by endogenous catecholamines: relationship of leukocyte beta receptor density to the cardiac sensitivity to isoproterenol. *J. Clin. Invest.* **67**:1777–1784 (1981).

21. I. M. Ramzan and G. Levy. Kinetics of drug action in disease states. XV: Effect of pregnancy on the convulsive activity of pentylenetetrazol in rats. *J. Pharm. Sci.* **74**:1233–1235 (1985).

PHARMACODYNAMIC MODELING OF THIOPENTAL DEPTH OF ANESTHESIA

Donald R. Stanski

Department of Anesthesia
Stanford University School of Medicine
and
Palo Alto Veterans Administration Medical Center

ABSTRACT

Quantitation of the degree of a central nervous depression in man necessary for surgical procedures can be achieved by measuring the anesthetic drug effect and using a pharmacodynamic modeling approach to relate drug concentration to drug effect. Using the intravenous anesthetic drug thiopental, EEG waveform analysis has been used to measure the CNS drug effect. The changes of EEG spectral edge or number of waves per second can be used to estimate CNS sensitivity. Using computer-controlled infusion pumps, constant thiopental plasma concentrations can be achieved and clinical responses (movement) observed from relevant noxious stimuli to define clinical anesthetic depth. Finally, it is also possible to relate the EEG measure of anesthetic drug effect to the clinically-observed measures of anesthetic depth.

INTRODUCTION

The clinical practice of anesthesia involves the selective depression of the central nervous system to create unconsciousness, analgesia, amnesia and muscle relaxation during surgical procedures. This is achieved by using intravenous anesthetics (e.g., thiopental) to rapidly induce unconsciousness, followed by a wide range of maintenance anesthetic drugs (e.g., isoflurane, nitrous oxide, opioids, muscle relaxants) for the remainder of the surgical procedure.

While many of the anesthetic drug effects can be readily measured (e.g., muscle paralysis, cardiovascular or respiratory depression), quantitation of the degree of central nervous system depression has eluded clinicians and scientists. The clinical dosing of anesthetic drugs to create CNS depression has remained one of the subjective *artforms* of anesthetic practice. Over the past 8 years, our research laboratory has pursued the quantitation and pharmacodynamic modeling of anesthetic drug effects on the CNS, using the intravenous induction agent thiopental.

Advanced Methods of Pharmacokinetic and Pharmacodynamic Systems Analysis
Edited by D'Argenio, Plenum Press, New York, 1991

METHODS AND RESULTS

Our initial efforts to quantitate the CNS drug effects of thiopental were directed to the electroencephalogram (EEG). This electrophysiological parameter was chosen because it is a continuous, noninvasive measure of CNS electrical activity that has a profound response to the administration of thiopental. Figure 1 displays the EEG changes induced with progressively increasing doses of thiopental. The EEG initially is activated (increase in amplitude and frequency) then progressively slows (decrease in frequency, increase in amplitude) until a burst/suppression pattern occurs which will progress to an isoelectric signal (flat EEG signal). Using fast fourier transform waveform analysis it is possible to quantitate the changes in EEG frequency and amplitude from which parameters can be chosen to distill the complex EEG signal into a single measure of the drug effect [1]. Initially we used the spectral edge, or EEG frequency below which 95% of the EEG power spectrum existed, as the measure of drug effect (Fig. 2). With the measurement of thiopental plasma concentrations and estimation of the EEG response, it was possible to use pharmacodynamic modeling concepts (effect compartment) to estimate brain responsiveness to thiopental [2,3]. The CP50 or steady state thiopental plasma concentration that resulted in one-half of the maximal EEG slowing was estimated as a measure of brain sensitivity.

This measure of CNS sensitivity was used to examine the effects of increasing age on thiopental pharmacokinetics and pharmacodynamics [4]. While the thiopental dose requirement declined 70% from age 20 to 80 years (Fig. 3), no change in brain sensitivity, as estimated by the EEG, was seen (Fig. 4). Altered thiopental distribution pharmacokinetics explained the age related decrease of thiopental dose requirement [5]. This same methodology was used to demonstrate that acute tolerance does not appear to exist with thiopental [3]. Also, chronic, heavy alcohol ingestion does not alter thiopental dose requirement or brain sensitivity [6].

One pharmacological limitation of the fourier analysis/spectral edge parameter is it's behavior as the EEG becomes isoelectric. When the EEG signal becomes extremely low in voltage, the fourier analysis is unable to quantitate the signal, and the spectral edge parameter no longer is a stable measure of the drug effect. The use of aperiodic analysis to estimate the number of EEG waves/sec allows quantitation of the EEG even when the EEG signal becomes isoelectric [7]. Figure 5 displays the thiopental steady state plasma concentration versus number of waves/sec relationship in six volunteer subjects [8]. At low thiopental concentrations, EEG activation occurs and the number of waves/sec increases. At high thiopental concentrations, the EEG slowing is characterized by the number of waves approaching zero.

While we have demonstrated the utility of the processed EEG as a pharma-

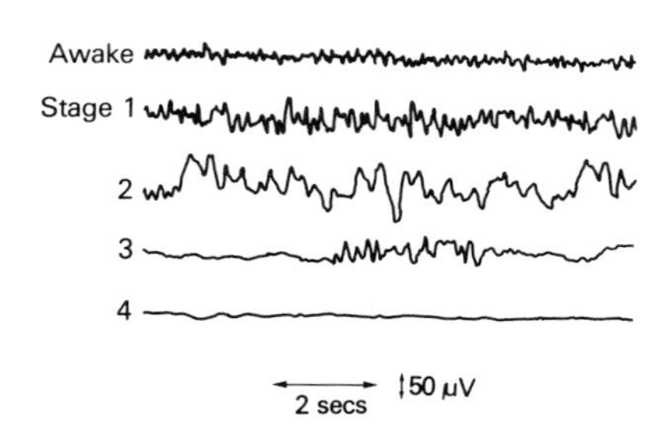

Fig. 1. EEG changes induced by thiopental. Note the biphasic EEG response with activation then the progressive infusion of slowing of the EEG signal.

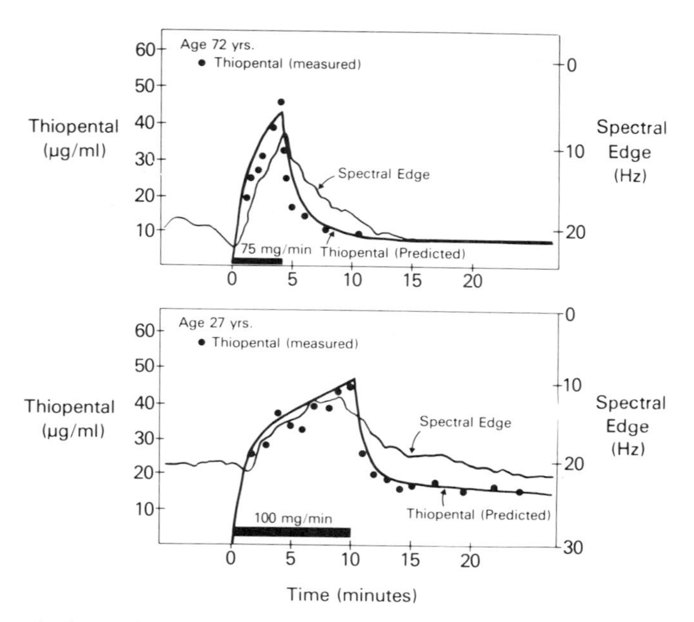

Fig. 2. The spectral edge and thiopental plasma concentration versus time curve in 2 patients, one young and one elderly. Note the lower thiopental dose requirement in the elderly patient.

cological tool, the relationship of thiopental's EEG changes to clinical measures of anesthetic depth were unknown. Estimation of clinical depth of anesthesia for thiopental proves to be much more complex than the EEG analysis. Clinical depth of anesthesia can only be assessed by examining the body's response (movement, blood pressure, pulse rate) to relevant noxious stimulation (electrical tetanus, muscle squeeze, laryngoscopy, intubation). To perform this assessment, one major pharmacokinetic hurdle must be overcome. Traditionally, thiopental is given as a rapid, intravenous bolus injection. This results in rapidly changing plasma and biophase concentrations as the thiopental is distributed/redistributed throughout the body. If a noxious stimulation occurs at the peak plasma or biophase concentrations, by the time the clinical response is manifest (30-60 sec later) drug concentrations will have changed markedly. To overcome this kinetic effervesce, it is necessary to "concentration clamp" the thiopental in plasma, allow the blood and CNS to equilibrate (3-5

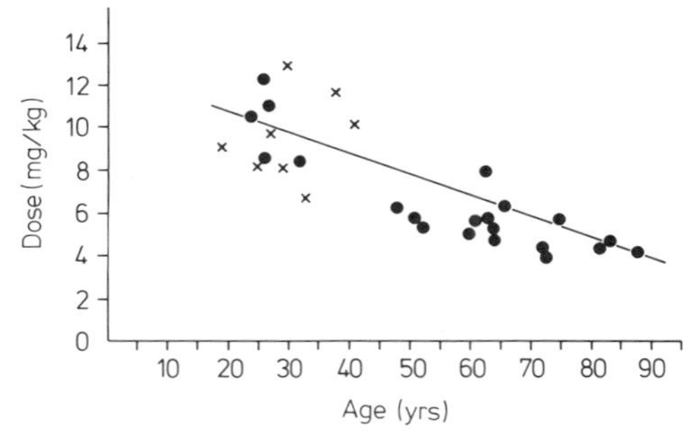

Fig. 3. Age-related decrease of thiopental dose requirement (dose needed to achieve burst suppression). x=volunteers, •=patients.

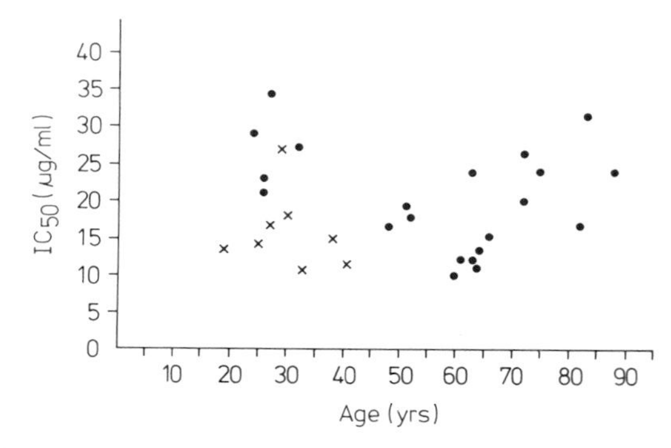

Fig. 4. The brain sensitivity IC50 (thiopental plasma concentration causing 1/2 of the maximal EEG slowing) estimated from pharmacodynamic modeling.

min) and then apply the relevant noxious stimuli. Because the plasma and biophase concentrations are not changing, one has ample time to both measure pharmacological phenomena and observe clinical responses. A computer driven infusion pump is needed to provide an exponentially declining infusion of thiopental to initially achieve, then maintain a constant blood concentration [9]. Figure 6 displays the ability of the computer driven infusion pump to maintain thiopental plasma concentrations at steps of 10 to 40 μg/ml for six minute periods of time in a volunteer subject [9]. It also demonstrates the EEG response in this subject using the number of waves/sec parameter.

To examine the relationship between thiopental plasma concentrations, EEG number of waves, clinical noxious stimulation and clinical measures of anesthetic depth, the following study was performed. Twenty six healthy, unpremedicated surgical patients were studied [10]. They ranged in age from 32 to 72 years. A radial artery catheter was used for continuous hemodynamic monitoring and frequent measurement of thiopental plasma concentrations and arterial blood gases. After awake, baseline hemodynamics and EEG were recorded, a low (10-30 μg/ml) stable thiopental concentration was achieved with a computer driven infusion pump. After five minutes at the stable concentration, the EEG number of waves had stabilized. The following stimuli were applied at one minute intervals: verbal responsiveness, electrical tetanus, trapezius muscle squeeze and direct laryngoscopy. Purposeful

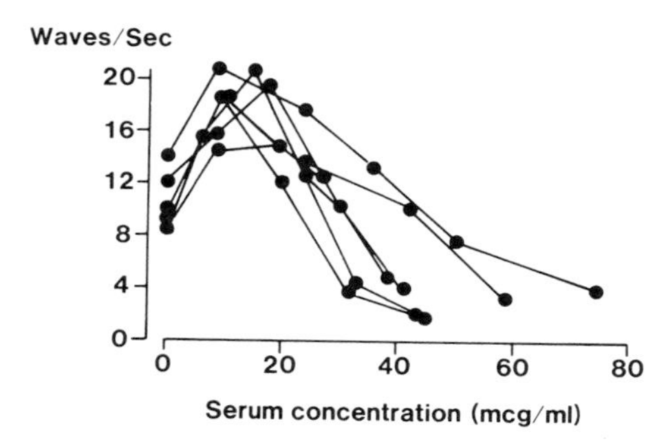

Fig. 5. The EEG waves/sec from aperiodic analysis versus steady-state thiopental plasma concentration in 6 volunteers. Note the biphasic concentration versus response relationship.

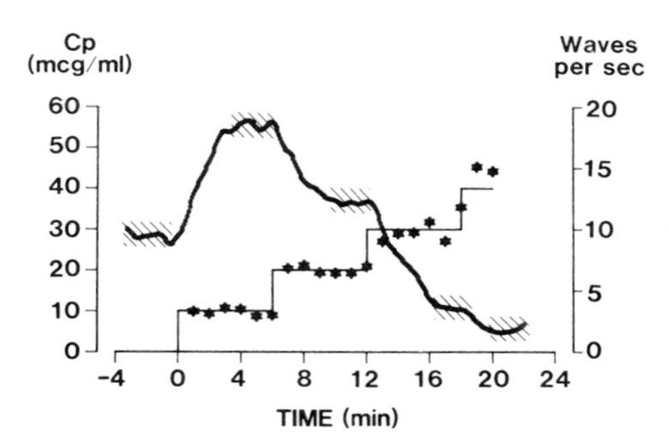

Fig. 6. Thiopental plasma concentration (✱) versus time from a computer-driven infusion pump and the EEG parameter waves/sec versus time. The /// is the equilibrated EEG effect at a constant thiopental plasma concentration.

movement, mean arterial blood pressure, heart rate and EEG response were recorded for each stimuli. A high (40-90 μg/ml) thiopental plasma concentration was then achieved and all of the stimuli repeated, including endotracheal intubation following a second direct laryngoscopy. Purposeful movement or a 15% increase in either mean arterial pressure or heart rate were considered positive clinical responses.

Figure 7 displays the move/no move versus thiopental plasma concentration relationship for the five different clinical stimuli. Figure 8 displays the logistic regression characterization of the data. The five different stimuli could be rank ordered in progressively increasing degrees of noxiousness as defined by higher thiopental CP50 values. A similar analysis could be performed for the hemodynamic variables. Meaningful pharmacodynamic relationships could only be estimated from the more noxious clinical stimuli: direct laryngoscopy and endotracheal intubation. Examining the relationship between the EEG number of waves and clinical measures of

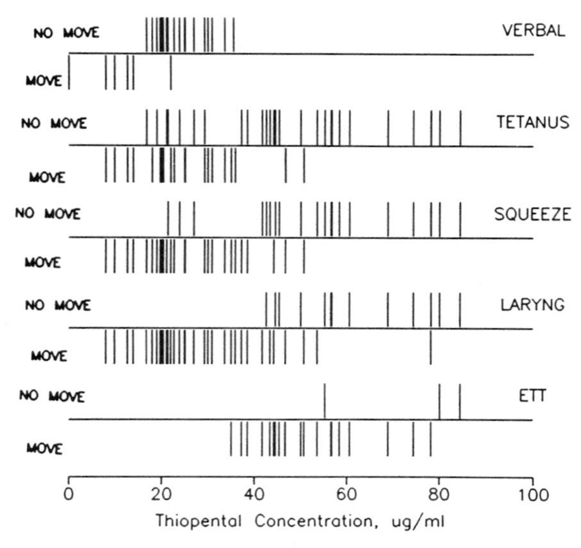

Fig. 7. The move/no move response at steady-state thiopental concentrations for different stimuli. Verbal is response to voice, tetanus is response to a 50 amp/5 sec electrical stimulus, squeeze is the response to trapezius muscle squeezing, laryngoscopy and intubation represent separate stimuli.

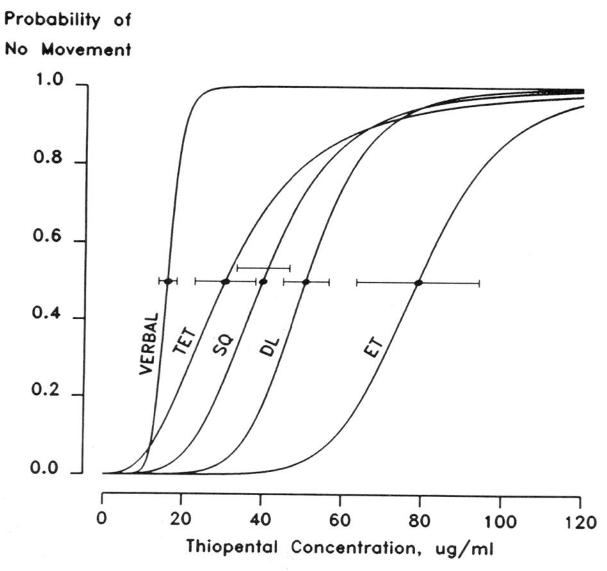

Fig. 8. Logistic regression of the data in Fig. 7, TET is tetanus, SQ is trapezius muscle squeeze, OL is laryngoscopy and ET is intubation.

thiopental anesthetic depth, unconsciousness occurred near the peak of EEG activation, or approximately 20 waves/sec. Suppression of movement response occurred at relatively high thiopental plasma concentrations that indicated profound EEG slowing. Zero to 5 waves/sec were required before movement responses were suppressed. The EEG signal did not increase (activate) even if the patient had profound clinical signs of inadequate thiopental anesthesia (movement, increase of hemodynamics).

DISCUSSION

The above studies demonstrate measurement of thiopental CNS drug effect using EEG electrophysiological and clinical measures. With the measurement of plasma thiopental concentrations, pharmacodynamic modeling could be used to link drug concentration to drug effect and allow estimation of CNS sensitivity. The measurement of clinical depth of anesthesia represented a greater technical and clinical challenge than the electrophysiological investigations. The use of computer-controlled infusion pumps to *concentration clamp* or *freeze* the drug plasma concentration was critical to performing the research where noxious stimulus and clinical response required a finite (1-3 minutes) period of time. It was possible to demonstrate how the EEG relates to clinical depth of anesthesia.

ACKNOWLEDGMENTS

Supported in part by the National Institute on Aging Grants P01-03104, R01-04594, The Veterans Administration Merit Review and The Anesthesiology/Pharmacology Research Foundation.

REFERENCES

1. W. Levy. Intraoperative EEG patterns: Implications for EEG monitoring. *Anesthesiology* **60**:430 (1984).
2. D. R. Stanski, R. R. Hudson, T. D. Homer, L. J. Saidman, and E. Meathe. Pharmacodynamic modeling of thiopental anesthesia. *J. Pharmacokin. Biopharm.* **12**:223-240 (1984).
3. R. J. Hudson, D. R. Stanski, E. Meathe, and L. J. Saidman. A model for studying depth of anesthesia and acute tolerance to thiopental. *Anesthesiology* **59**:301-308 (1983).
4. T. D. Homer and D. R. Stanski. The effect of increasing age on thiopental disposition and anesthetic requirement. *Anesthesiology* **62**:714-724 (1985).
5. D. R. Stanski and P. O. Maitre. Population pharmacokinetics and pharmacodynamics of thiopental: the effect of age revisited. *Anesthesiology* **72**:412-422 (1990).
6. B. N. Swerdlow, F. O. Holley, P. O. Maitre, and D. R. Stanski. Chronic alcohol intake does not change thiopental anesthetic requirement, pharmacokinetics or pharmacodynamics. *Anesthesiology* **72**:455-461 (1990).
7. T. K. Gregory and D. C. Pettus. An electroencephalographic processing algorithm specifically intended for analysis of cerebral electrical activity. *J. Clin. Monit.* **2**:190 (1986).
8. M. Bührer, P. O. Maitre, W. F. Ebling, and M. D. Stanski. Defining thiopental's steady state plasma concentration - EEG effect relationship. *Anesthesiology* **67**:A399 (1987).
9. S. L. Shafer, L. C. Siegel, J. E. Cooke, and J. C. Scott. Testing computer-controlled infusion pumps by simulation. *Anesthesiology* **68**:261-266 (1988).
10. O. R. Hung, J. R. Varvel, S. L. Shafer, and D. R. Stanski. The use of computer-controlled infusion pump to study anesthetic depth. *Clin. Pharm. Ther.* **47**:209 (1990).

PHARMACOMETRICS:
MODELING, ESTIMATION AND CONTROL

RESIDENCE TIME DISTRIBUTIONS IN PHARMACOKINETICS: BEHAVIORAL AND STRUCTURAL MODELS

Michael Weiss

Institut für Pharmakologie und Toxikologie
Martin-Luther-Universität Halle-Wittenberg

INTRODUCTION

In contrast to other mathematical concepts used in pharmacokinetics the theory of residence time distributions (RTDs) is independent of a detailed structural model or a particular curve model. In view of the fact that interpretations of RTDs have been mostly based on compartmental models (e.g., [1,2]), the following shortcomings of this class of structural models should be noted: 1) there is no *a priori* reason for the existence of homogeneous compartments. The assumption that all elementary subsystems are characterized by exponentially distributed transit times restricts the generality of the approach; 2) the definition of a sampling compartment from which elimination occurs does not allow for a differentiation between sampling upstream or downstream of the elimination site, which leads to inconsistencies in the definition of the volume of distribution, V_{ss} [3].

This chapter reviews results based on the concept of RTDs aiming at a unified view of linear pharmacokinetics. The fact that the physical basis of pharmacokinetics is given by the transport processes of drugs in the body, necessarily implies a circulatory structure. Since the residence time of a molecule is defined as its age at its departure from the system, however, an additional assumption has to be made if the observation does not take place at the system output.

In what follows the term behavioral models will be used to refer to models which are based on the *a posteriori* evidence of a certain curve shape (e.g., log-convexity of disposition curves) or monotonicity properties of characteristic functions. The corresponding concept of nonparametric classes of RTDs allows an interpretation of the role of RTD moments (mean and variance of RTDs), and generalizes previous results obtained for empirical curve models (or parametric families of RTDs) like a sum of exponential functions.

The probability description used in this chapter – most results are based on theorems of probability theory (e.g., [4–6]) – does not mean that its application would be restricted to probabilistic systems; although the processes are stochastic at a microscopic (molecular) level, the assumption that all macroscopic quantities or observables are in principle measurable with any desired accuracy implies that the pharmacokinetic system is deterministic at the macro level.

Advanced Methods of Pharmacokinetic and Pharmacodynamic Systems Analysis
Edited by D'Argenio, Plenum Press, New York, 1991

RESIDENCE TIME DISTRIBUTIONS

The *residence time*, T, of a molecule in a system is a positive random variable with cumulative RTD

$$F(t) = P(T \leq t) \tag{1}$$

which is defined as the probability that a molecule has a residence time less than or equal to t. (Although *residence* time is the established term, *exit* time would be more appropriate.) Suppose $F(t)$ is absolutely continuous on $[0, \infty)$, its first derivative $f(t) = dF/dt$ is called RTD density. Then $f(t)dt$ represents the probability that the residence time is in the interval $(t, t+dt]$. This implies that $F(t)$ must be nondecreasing with $F(0-) = 0$ and $F(\infty) = 1$. Let $\overline{F}(t) = 1 - F(t) = P(T > t)$ (survivor function). The failure (or intensity) rate function defined according to

$$k(t) = f(t)/\overline{F}(t)) \tag{2}$$

is a conditional probability density function: $k(t)dt$ represents the probability that a molecule *of age* t will leave the system in the interval $(t, t + dt]$. The advantage of the function $k(t)$ is its sensitivity to certain properties of RTDs. As noted above, $F(t)$ is in fact the distribution of ages of molecules when they leave the system (exit age distribution). In contrast, the *internal age* of a molecule, T_A, (i.e., the current life along its trajectory through the system) with distribution $G(t)$ has a density

$$g(t) = \overline{F}(t)/E(T) \tag{3}$$

where $E(T)$ is the expected value of the exit age T (mean residence time).

RTDs can be characterized by their moments

$$MO_n = \int_0^\infty t^n f(t)dt = \int_0^\infty t^n dF(t) = n \int_0^\infty t^{n-1}\overline{F}(t)dt \tag{4}$$

which determine the expectation, $E(T_n) = MO_n$. Thus, the mean and variance of RTD are given by

$$MRT = MO_1 \tag{5}$$

$$VRT = MO_2 - MO_1 \tag{6}$$

The dimensionless variance

$$CV^2 = VRT/MRT^2 \tag{7}$$

is also called *relative dispersion* of RTDs. CV^2 and normalized higher moments provide dimensionless measures of the shape of RTDs.

In order to define $f(t)$ in terms of observables we shift from individual molecules (micro level) to large ensembles of molecules (macro level) and make the following assumption:

A1. The molecules are indistinguishable, do not interact with each other and their individual residence times are mutually independent. A molecule which left the system will never return.

Then it follows from the above definition that $f(t)dt$ represents the fraction of the population of molecules – which have entered the system at the same time – for which the residence time is in $(t, t+dt]$, i.e., the ratio of the amount dA_e which leaves the system (is eliminated) in this interval to the total amount eliminated in all time. Consequently, we have

$$f(t) = (dA_e/dt)/A_e(\infty) \tag{8}$$

$$F(t) = A_e(t)/Ae(\infty) \tag{9}$$

$$\overline{F}(t) = ARE(t)/A_e(\infty) \tag{10}$$

where $ARE(t) = A_e(\infty) - A_e(t)$ denotes the amount remaining to be eliminated and $k(t)$ can be interpreted as the *fractional elimination rate*

$$k(t) = (dA_e/dt)/ARE(t) \tag{11}$$

For a single-pass flow system we suppose that the molecules are transported by a fluid flow with flow rate Q, we have $dA_e/dt = QC_{\text{out}}$, and $f(t)$ is completely determined by the outlet concentration, C_{out}. For the whole body system (recirculatory system) the following assumption is made:

A2. The elimination rate is proportional to the sampled concentration: $dA_e/dt = CL \cdot C(t)$.

Note that A2 represents the definition of the clearance, CL, and implies instantaneous elimination at an elimination site just downstream to the point of observation. For arterial sampling this appears to be in agreement with available experimental data if pulmonary elimination can be neglected.

From Eq. (8) and Assumption A2 it follows that the exit time density $f(t)$ can be estimated from measured concentration-time curves

$$f(t) = C(t)/AUC \tag{12}$$

where $AUC = \int_0^\infty C(t)dt = A_e(\infty)/CL$.

Concentration-time curves or RTDs after impulse input (bolus intravenous injection) are called *disposition curves* or *disposition RTDs*, respectively, and $f_D(t)$ is determined by $C_D(t)$ (Eq. (12)), the unit impulse response of the system. Consequently, the cumulative disposition RTD, $F_D(t)$, is identical to the normalized accumulation curve following constant-rate infusion (positive step change of input), $C_A(t)/C_{ss}$, or the time course of cumulative urinary excretion $A_{e,R}(t)/A_{e,R}(\infty)$. It is particularly interesting that the *washout curve* following termination of infusion (negative step change of input), $C_W(t)$, determines the survivor function

$$\overline{F}_D(t) = C_W(t)/C_{ss} \tag{13}$$

The mean, $E(T_D) = MDRT$, is called *mean disposition residence time.*

For the sake of a clear notation the index B will be used to characterize concentration-time curves or RTDs following noninstantaneous input (examples are oral and transdermal administration), with $f_B(t)$ given by Eq. (12) and $F_B(t)$ by Eq. (9).

NONPARAMETRIC CLASSES OF RESIDENCE TIME DISTRIBUTIONS

We define three complementary nonparametric classes of RTDs:

D1. A RTD F belongs to the DFR (IFR) class if $k(t)$ is monotone decreasing (increasing) and to the NMFR class if $k(t)$ is nonmonotonic.

For the monotonic classes the following theorem was proved [4]:

T1.

a. The log-convexity (log-concavity) of $f(t)$ implies the DFR (IFR) property.
b. F is DFR (IFR) if and only if $F(t)$ is log-convex (log-concave).

With regard to the asymptotic properties of RTDs we define the class $S(\gamma)$ according to [7]:

D2. A RTD F belongs to the $S(\gamma)$ class if and only if

a. $\lim_{t\to\infty}(1 - F * F)/\overline{F} = 2\int_0^\infty e^{-\gamma t}dF(t) < \infty$
b. $\lim_{t\to\infty} \overline{F}(t+x)/\overline{F}(t) = e^{-\gamma x}$

Then the following theorem can be proved:

T2. The RTD of a pharmacokinetic system belongs to the class $S(\gamma)$, which implies the asymptotic properties $\lim_{t\to\infty} k(t) = \gamma$ and consequently $F(t) \approx c_F e^{-\gamma t}$ and $G(t) \approx c_G e^{-\gamma t}$ for t sufficiently large.

The proof is based on the fact that any molecule eventually leaves an open system, i.e., according to the laws of thermodynamics the conditional mean residual life of a molecule at age t, $E(T_R - t|T_R > t)$, cannot be infinite. Consequently, since $k_G(t) = (E(T_R - t|TR > t))^{-1}$ we have $k_G(\infty) > 0$. From $\lim_{t\to\infty} k_G(t) = \gamma$ it follows that G belongs to $S(\gamma)$ [7] and since $\lim_{t\to\infty} k(t) = \lim_{t\to\infty} k_G(t)$ [6], the asymptotic properties of $F(t)$ and $G(t)$ follow.

DFR Class

From Eq. (12) and Theorem T1b it follows that the log-convexity of washout curves ($logC_W(t)$ convex) is a necessary condition for the DFR property of RTDs, and we can infer that the DFR class is the natural class of disposition RTDs. If we deal with log-convex $C_D(t)$ curves the following assumption is made:

A3. The brief initial rise (concavity) of $C_D(t)$ observed immediately after impulse injection can be neglected assuming that the underlying process of initial distribution is sufficiently rapid compared to the time scale of the experiment.

There is experimental evidence that this assumption is reasonable; in humans, for example, decreasing $C_D(t)$ curves have been observed for $t > 1$ min [8].

Properties of systems with RTDs of the DFR class allow generalizations in linear pharmacokinetics [9]:

T3. If the disposition curve, $C_D(t)$, and/or the washout curve, $C_W(t)$, are log-convex, then

a. The tails of washout curves and disposition curves are asymptotically exponentials, i.e., $C_W(t) \approx c_w e^{-\lambda_Z t}$ and $C_D(t) \approx c_D e^{-\lambda_Z t}$ (t sufficiently large), where $\lambda_Z = \lim_{t\to\infty} k_D(t)$.

b. The dynamic volume of distribution defined as $V(t) = ARE_D(t)/C_D(t)$ is monotone increasing until it approaches its asymptotic value $V_Z = CL/\lambda_Z$.

c. $CV_D^2 \geq 1$ and the scale invariant parameter $d_D = (CV_D^2 - 1)/2$ is a measure of departure from the well-mixed system with monoexponential RTD (characterized by $CV_D^2 = 1$ and $d_D = 0$).

The properties (a) and (b) follow from Theorem T2 recalling that $V(t) = CL/k_D(t)$ with $k_D(t)$ increasing. The result (c) has been proved for DFR distributions by Brown [6]. Only for the well-mixed system (one-compartment model) is the age distribution, $G(t)$, identical to the RTD (exit age distribution), FD(t), since then the age of a molecule leaving the system is identical to the age of any molecule at any point within the system.

IFR Class and NMFR Class

The drug enters the system via a noninstantaneous input process if the observed concentration-time curve is unimodal instead of monotone decreasing. The corresponding RTDs then belong to the NMFR or IFR class. For the latter one has the following results [10]:

T4. If the concentration-time curve, $C_B(t)$, is log-concave then

a. The fractional elimination rate $k_B(t)$ is increasing in t until it reaches its asymptotic value $k_{B,Z} \geq 1/MBRT$.

b. The $C_B(t)$ curve has an exponential tail, $C_B(t) \approx c_B e^{-k_{B,Z} t}$ for t sufficiently large.

c. $CV_B^2 < 1$

For F IFR (cf. T1) the result (c) is well-known [4]; to prove (a) and (b), note that $k_G(t) \approx 1/MBRT$ and $\sup k(t) \geq k_G(t)$ [11].

The fact that the total random residence time, T_B, of a molecule with an *input time*, T_I, into the disposition system is the sum, $T_B = T_I + T_D$, implies $F_B = F_I * F_D$, where $*$ denotes the convolution operation. Obviously, the minimum structural model for the IFR and NMFR classes of RTDs in pharmacokinetics consists of two consecutive subsystems with distributions F_I and F_D, respectively. Consequently, we have $MBRT = MIT + MDRT$ and $VBRT = VIT + VDRT$. Based on the inequality $CV_B^2 < 1$ (cf. T3c) we can derive necessary (but not sufficient) conditions for the generation of log-concave $C_B(t)$ curves. Defining $d_I = (CV_I^2 - 1)/2$, where CV_I^2 is the relative dispersion of the input time distribution, we obtain the following bounds on the ratio mean input time to mean disposition residence time, $MIT/MDRT$:

$$MIT/MDRT > 2/d_I + \sqrt{4/d_I^2 + d_D/d_I}, \qquad \text{for } d_I < 0 \tag{14}$$

$$MIT/MDRT > d_D, \qquad \text{for } d_I = 0 \tag{15}$$

In words, for a given departure of F_I and F_D from the exponential distribution (as measured by d_I and d_D, respectively) it depends on *MIT* (scaled by *MDRT*) whether the resulting $C_B(t)$ curve is log-concave. Thus, for $CV_I^2 \leq 1$ (F_I IFR) the threshold value of $MIT/MDRT$ for the generation of log-concave $C_B(t)$ curves increases if the rate of drug distribution in the body decreases. Within the classical pharmacokinetic concept of polyexponential $C_B(t)$ curves this phenomenon is well-known as that of *vanishing of exponential terms.* Obviously, log-concavity is dependent on both the rate of distribution and the input rate. When dissolution represents the rate-limiting step of the input process after oral administration, it has been conjectured that then $F_I \approx F_{\text{Diss}}$ is IFR and $k_{B,Z} \approx k_{\text{Diss},Z}$ (if $k_{\text{Diss},Z} \ll k_{D,Z}$) [10].

For RTDs of the NMFR class (with unimodal density) the function $k_B(t)$ is also unimodal with zero initial value and asymptotic value $k_{B,Z}$ (cf. Theorem T2). From Eqs. (10) and (11) it follows that $ARE(t)$ is log-concave (log-convex) when $k_B(t)$ is increasing (decreasing), i.e., for $t < t_{max}$ $(t > t_{max})$ where t_{max} is the mode point of $k_B(t)$. Recall that for the DFR (IFR) class $ARE(t)$ is log-convex (log-concave). As discussed above, the *distributional nose* of $C_B(t)$ curves – as a characteristic of the NMFR property – may disappear if $MAT/MDRT$ increases for $CV_I^2 \leq 1$.

PARAMETRIC FAMILIES OF RESIDENCE TIME DISTRIBUTIONS

DFR Class

A RTD belongs to the family CM (completely monotone) if all derivatives of f_D exist and $(-1)^n f_D^{(n)}(t) > 0, n = 1, 2, \ldots$ [12]. The CM family is of great importance to pharmacokinetics since f_D is in this case a mixture of exponential densities, i.e.,

$$C_d(t) = \int_0^\infty e^{-\lambda_t} s(\lambda) d\lambda \tag{16}$$

and under the assumption of a discrete spectrum, $s(\lambda) = \sum_i c_i \delta(\lambda - \lambda_i)$, one obtains a polyexponential drug disposition curve which represents the commonly used disposition function and arises naturally from compartmental models.

As an alternative to hyperexponential RTDs the use of gamma distributed disposition residence times has been proposed [13,14]:

$$C_D(t) = Ae^{-a}e^{-bt}, \quad \begin{cases} 0 < a < 1 \\ b > 0 \\ t > 0 \end{cases} \tag{17}$$

Despite the obvious advantages of gamma curves, such as fewer parameters and the fact that the shape parameter, a, completely determines $CV_D^2 = 1/(1-a)$ (b is a scale parameter), the superiority of polyexponential functions has been demonstrated for amiodarone where gamma curves showed systematic deviations from the measured data [15].

It should be emphasized that the Weibull distribution (with $\gamma = 0$) which was recently proposed as a disposition RTD model [16] is not compatible with the physical behavior of molecules (see Theorem T2). For the same reason, residence times cannot be lognormal distributed – as sometimes assumed (e.g., [17]) – since the RTDs then would belong to the class $S(0)$, called subexponential distributed [7], and not to $S(\gamma)$.

IFR Class

The most popular parametric family has a density which is proportional to

$$C_B(t) = A(e^{-\lambda_1 t} + e^{-\lambda_2 t}), \qquad \text{for } 1/2 < CV_B^2 < 1 \tag{18}$$

which is called Bateman function. Note that $k_{B,Z} = min(\lambda_1, \lambda_2)$. Although Eq. (18) results if both F_I and F_D are exponential distributions, this conclusion is not reversible; consequently, from a good fit with a Bateman function one cannot infer that the disposition process is monoexponential (one-compartment model).

Unimodal gamma curves which were introduced into pharmacokinetics by van Rossum and van Ginneken [18],

$$C_B(t) = At^a e^{-bt}, \qquad a > 0 \tag{19}$$

have important advantages. There is no restriction for CV_B^2 (as in Eq. (18)) and $CV_B^2 = 1/(1+a)$ determines the mode point t_{max} according to $t_{max} = MBRT$, $d_B = MBRT(1 - CV_B^2)$ (13).

NMFR Class

If the weights F_i in the mixture of exponential RTDs

$$F_B(t) = \sum_{i=1}^{n} F_i(1 - e^{-\lambda_i t}), \qquad \sum_{i=1}^{n} F_i = 1 \tag{20}$$

are allowed to be negative any $C_B(t)$ curve can be approximated arbitrarily closely by the density of F_B

$$C_B(t) = \sum_{i=1}^{n} C_i e^{-\lambda_i t}, \qquad \sum_{i=1}^{n} C_i = F_i \lambda_i / AUC \tag{21}$$

A classical example is the two-compartment model with first-order absorption. First-order input, however, is an exception, and practical difficulties in curve fitting may arise for $n > 3$. Thus, the following alternative concept should be of interest.

A unimodal (but not concave) density function which fulfills Theorem T2 is that of the generalized inverse Gaussian distribution (GIGD)

$$f_B(t) = \frac{(\psi/\chi)^{\lambda/2}}{2K_\lambda(\sqrt{\chi\psi})} t^{\lambda-1} e^{-(\chi t^{-1} + \psi t)/2}, \qquad \lambda < 1, \quad \psi > 0 \tag{22}$$

where K_λ is the modified Bessel function of the third kind with index λ. The mode point is given by

$$t_{max} = \left(\lambda - 1 + \sqrt{(\lambda - 1)^2 + \chi\psi}\right) / \psi \tag{23}$$

and $k_B(t)$ has an asymptotic value $k_{B,Z} = \psi/2$ (for further results see [19]). One special case is the inverse Gaussian distribution ($\psi = 0, \lambda = -1/2$), which may be written as

$$f_B(t) = \frac{a}{\sqrt{2\pi}\, t^{3/2}}\, e^{-(a-bt)^2/2t} \tag{24}$$

where $MBRT = a/b, CV_B^2 = 1/(ab)$ and $k_{B,Z} = b^2/2 = 1/(2CV_B^2\, MBRT)$. From a physical point of view the role of these functions as first passage densities of Wiener processes with drift suggests their potential use as residence time distributions. While applications to $C_B(t)$ data are yet missing, Eq. (24) was successfully fitted to impulse response curves of the isolated perfused rat liver, i.e., the inverse Gaussian distribution appears to be an appropriate empirical model for liver RTD [20].

CIRCULATORY STRUCTURE

Since convective circulatory transport is the basic physical process of disposition kinetics (primarily convective transport to the organs and diffusion within the organ) the assumption of a circulatory structure does not reduce the generality of the work. The *circulation* or *cycle time*, T_C, of a molecule is defined as the time between its consecutive passages through a reference cross-section, and the *cycle time distribution* CTD is the RTD of the single-pass system S_C (F_C) which results if the circulatory system S_D (with RTD F_D) is assumed to be cut open at the reference cross-section. Consequently, we have to define the points of injection and sampling in S_D with respect to the reference cross-section.

A4. The points of observation (blood sampling) and injection (bolus dose D) are located just downstream and upstream, respectively, to the reference cross-section.

Arterial sampling and intravenous injection is in accordance with Assumptions A2 and A4 if the influence of the lungs is neglected; the approach is generally valid, however, if sampling occurs before the lungs.

Let $T_{C0} = 0$ and $T_{C1}, T_{C2}, \ldots$ be a sequence of independent, identically distributed (i.i.d.) random variables with distribution F_C (successive recirculations of a molecule after its first passage (T_{C0}) at the reference point). This process is well known in probability theory as renewal process [5,21]. The residence time of a molecule until the occurrence of the nth recirculation is then given by

$$S_n = \sum_{i=0}^{n} T_{Ci}, \qquad S_0 = 0 \tag{25}$$

T_{Ci} are i.i.d., the distribution function $F_n = P(S_n \leq t)$ is the n-fold convolution of F_C with itself, denoted by $F_n = F(n)$.

Noneliminating System

In order to analyze drug distribution in the body unaffected by elimination we study the spontaneous evolution of the closed (noneliminating) system S^* towards an equilibrium state. Let $N(t)$ be the random number of recirculations in $(0, t]$, i.e., $N(t) = max(i : S_i \leq t)$, then we have $P[N(t) \geq n] = P(S_n \leq t)$. The expected number of recirculations is given by

$$E[N(t)] = \sum_{i=1}^{\infty} P[(t) \geq i] = \sum_{i=1}^{\infty} F_{Ci}(t) \tag{26}$$

Since according to the above definition the event with T_{C0} is not counted, we include this first event in order to define the number of passages at the reference point $M(t)$ by

$$M(t) = E[N(t)+1] = \sum_{i=0}^{\infty} F_{Ci}(t) \tag{27}$$

Then $m(t) = dM/dt$ is the probability of a passage in $(t, t+dt]$, and at the macro level $m(t)dt$ can be interpreted as the fraction of mass, dA_r/D, which passes the reference point in $(t, t+dt]$. Since $dA_r = QC^*(t)dt$ (where Q denotes total blood flow and $C^*(t)$ is the concentration profile observed in S^*) we obtain

$$m(t) = QC^*(t)/D \tag{28}$$

For the state of complete equilibrium the following holds.

T5.

a. Let $MCT = E(T_C) = \int_0^\infty \overline{F}_C(t)dt$ be the *mean circulation time*, then: $\lim C^*(t) = C^*(\infty) = D/(Q \cdot MCT)$.

b. The *volume of distribution* of S^* defined by $V_{ss} = D/C^*(\infty)$ is determined by the equilibrium partition coefficients K_i of the various phases with anatomical tissue volumes V_i, according to $V_{ss} = Q \cdot MCT = \sum_i K_i V_i$.

The result (a) follows directly from the renewal density theorem [5,22] and (b) is a consequence of the fact that in a closed multiphase system all phases are in equilibrium at steady state: we have $D = \sum_i V_i C_i$ and obtain (b) substituting $K_i = C_i/C_1$ (C_1 is the reference concentration).

In order to characterize the transient behavior, i.e., the dynamics of drug distribution, we introduce the concept of a *mixing curve*, $C_M(t)$, and the area under the mixing curve, AUC_M, according to $C_M(t) = C^*(t) - C^*(\infty)$ and $AUC_M = \int_0^\infty C_M(t)dt$, respectively. Note that AUC_M increases with increasing departure of the system from the well-mixed behavior, i.e., from the case of instantaneous distribution after bolus injection where $C_M(t) = C_{M,\delta} = (D/Q)\delta(t)$ and $AUC_{M,\delta} = D/Q$ (where $\delta(t)$ denotes the impulse function). The following results were proved for CTDs of the DFR class.

T6.

a. If F_C is DFR, then $C^*(t)$ is monotone decreasing until $C^*(\infty)$ is attained [22,23].

b. If F_C is DFR, then the scaled distance between AUC_M and $AUC_{M,\delta}$ is given by [24]:

$$(AUC_M - AUC_{M,\delta})/AUC_{M,\delta} = (CV_C^2 - 1)/2 = d_C \tag{29}$$

Thus, the relative dispersion of CTDs, CV_C^2, quantifies the mixing behavior, i.e., the transient departure of the system from equilibrium distribution, while MCT determines the extent of distribution at equilibrium. Obviously, instantaneous distribution following bolus injection (well-mixed system) is obtained for exponentially distributed cycle times ($CV_C^2 = 1, dC = 0$). Note the analogy between the role of CV_D^2 and CV_C^2 as measures of the dynamics of drug distribution in the open and closed system, respectively (cf. Theorem T3c). Systems with monotone decreasing mixing curves have been termed passive [22].

Eliminating System

Since any molecule is eventually eliminated, the residence time, T_D, of a molecule is the sum of its N_e cycle times,

$$T_D = \sum_{i=0}^{N_e} T_{Ci} \tag{30}$$

The total number, N_e, of recirculations in all time is geometrically distributed

$$P(N_e \geq i) = (1 - E)^i, \qquad 0 < E < 1, \quad i = 0, 1, 2, \ldots \tag{31}$$

when $(1 - E)$ is the probability that a molecule will survive a single passage, and E is the probability that a molecule will be eliminated during one cycle. The expected number of recirculations is then given by

$$E(N_e) \;=\; \sum_{i=0}^{\infty} P(N_e \geq 1) \;=\; 1/E - 1 \tag{32}$$

Using Wald's identity (e.g., [4]) we obtain from Eq. (30):

$$MDRT \;=\; E(T_D) \;=\; Ed(T_C)E(N_e) \;=\; MCT(1/E - 1) \tag{33}$$

The relationship between higher moments of RTD and CTD is also dependent on E [24]. It can be readily shown that at the macro level E is identical to the single-pass extraction ratio $E = CL/Q$. Note that Eq. (33) implies, $MDRTCL = (1 - E)V_{ss}^*$ [3].

T7. If F_C belongs to the DFR class, then F_D belongs to the DFR class [22].

Based on Theorem T7 and the conjecture that decreasing $m(t)$ (cf. T6a) is also a necessary condition for the DFR property of the corresponding distribution [25], one could infer that F_D DFR is an intrinsic property of passive systems. Consequently, the significance of the DFR class of disposition RTDs can be explained on the one hand by the *a posteriori* evidence of log-convexity and one the other hand via an interpretation of the second law of thermodynamics (evolution of the system towards an equilibrium state).

Interestingly, it could be shown that the assumption of a GIGD for the CTD (i.e., application of Eq. (21) as $f_C(t)$ model) leads for low extracted drugs ($E \ll 1$) to a gamma disposition RTD see Eq. (16) [26].

Multiorgan Structure

To gain an intuitive insight into the physiological factors determining CV_C^2 (i.e., the dynamics of drug distribution in the closed system) we have to adopt a more detailed structural model of the underlying transport processes, such as physiological flow models. (The mean MCT, in contrast, is solely dependent on anatomical and physicochemical parameters, see Theorem T5b). To simplify the representation we neglect the role of the lungs and assume that the n organs of the systemic circulation with perfusion rates Q_i are arranged in parallel. The mean and variance of CTD is then given by [27]

$$MCT = \sum_{i=1}^{n}(Q_i/Q)MTT_i \tag{34}$$

$$VCT = \sum_{i=1}^{n}(Q_i/Q)(MTT_i - MCT)^2 + \sum_{i=1}^{n}(Q_i/Q)VTT_i \tag{35}$$

where MTT_i and VTT_i are the means and variances of *organ transit time*. Note that $\sum_i Q_i = Q$ and $MTT_i = V_i^*/Q_i$, where V_i^* denotes the distribution volume of ith subsystem at equilibrium.

While MCT is simply the (flow-weighted) average of the mean organ transit times, VCT is determined both by the processes of drug distribution *among* and *within* subsystems as indicated by the first and second term in Eq. (35), respectively. As noted above for CV_C^2, we need a geometrical model of the ith organ to predict CV_i^2. For the simplest model, based on the assumption of instantaneous distribution within the organs (one-compartment organ model), one obtains [13]

$$d_C = (CV_C^2 - 1)/2 = \sum_{i=1}^{n}(V_i^*/V_{ss}^*)^2/(Q_i/Q) - 1 \tag{36}$$

which represents a measure of the heterogeneity of mean organ transit times (coefficient of variation of the MTT_i). (Note that $V_i^* = K_iV_i$, cf. Theorem T5b). The dynamics of distribution (and consequently the quantity AUC_M, cf. T6b) is, therefore, affected by changes in hemodynamics, e.g., by a redistribution of cardiac output. The simplest organ model which accounts for diffusive tissue transport is the two-compartment organ model where CV_i^2 increases with decreasing membrane permeability [27]. Since DFR organ transit time distributions imply F_D DFR (mixtures of DFR distributions are DFR [4]), the DFR property of disposition RTDs is in accordance with the observed log-convexity of organ washout curves. Concepts based on IFR subsystems [1], on the other hand, are not applicable to such physiological multiorgan models.

CONCLUDING REMARKS

Nonparametric classes of RTDs and the concept of drug recirculation (renewal theory) provide a robust basis for pharmacokinetic modeling. We have seen that the DFR class plays a fundamental role for both the disposition residence time and circulation time distribution of drugs, where the relative dispersion acts as a measure of the departure of the system from the well-mixed behavior (rate of distribution). An input system in series with the disposition system then generates RTDs of either the NMFR or IFR class, respectively. One application of the DFR and IFR property in therapeutics and toxicokinetics is the prediction of washout and accumulation periods of drugs [28,29].

There are, of course, no models without simplifications; here, we have neglected (i) nonmonotonicity due to the initial distribution process (Assumption A3) and (ii) the influence of a delayed distribution within the eliminating organs (Assumption A2). The present approach, therefore, does not account for the peak and the oscillations of blood concentration profile observed within the first minutes after bolus injection [30]. The validity of the clearance concept (i.e., of Assumption A3) is well accepted in pharmacokinetics; at least it represents a useful approximation as long as it has not been disproved experimentally.

REFERENCES

1. J. H. Matis, T. E. Wehrly, and C. M. Metzler. On some stochastic formulations and related statistical moments of pharmacokinetic models. *J. Pharmacokin. Biopharm.* **11**:77-92 (1983).
2. S. L. Beal. Some clarifications regarding moments of residence times with pharmacokinetic models. *J. Pharmacokin. Biopharm.* **15**:75-92 (1987).
3. M. Weiss. Nonidentity of the steady-state volumes of distribution of the eliminating and noneliminating system. *J. Pharm. Sci.* (in press).
4. R. E. Barlow and F. Proschan. *Statistical Theory of Reliability and Life Testing*, Holt, Rinehart and Winston, New York, 1975.
5. W. Feller. *An Introduction to Probability Theory and its Applications*, Vol. 2, Wiley, New York, 1966.
6. M. Brown. Approximating IMRL distributions by exponential distributions, with applications to first passage times. *Ann. Probab.* **11**:419-427 (1983).
7. P. Embrechts. A property of the generalized inverse Gaussian distribution with some applications. *J. Appl. Probab.* **20**:537-544 (1983).
8. T. K. Henthorn, M. J. Avram, and T. C. Krejcie. Intravascular mixing and drug distribution: The concurrent disposition of thiopental and indocyanine green. *Clin. Pharmacol. Ther.* **45**:56-65 (1989).
9. M. Weiss. Generalizations in linear pharmacokinetics using properties of certain classes of residence time distributions. I. Log-convex drug disposition curves. *J. Pharmacokin. Biopharm.* **14**:635-657 (1986).
10. M. Weiss. Generalizations in linear pharmacokinetics using properties of certain classes of residence time distributions. II. Log-concave concentration-time curves following oral administration. *J. Pharmacokin. Biopharm.* **15**:57-74 (1987).
11. M. Brown and G. Ge. Exponential approximations for two classes of aging distributions. *Ann. Probab.* **12**:869-875 (1984).
12. J. Keilson. *Markov Chain Models - Rarity and Exponentiality*, Springer-Verlag, New York, 1979.
13. M. Weiss. Use of gamma distributed residence times in pharmacokinetics. *Eur. J. Clin. Pharmacol.* **25**:695-702 (1983).
14. M. E. Wise. Negative power functions of time in pharmacokinetics and their implications. *J. Pharmacokin. Biopharm.* **13**:309-346 (1985).
15. G. T. Tucker, P. R. Jackson, G. C. A. Storey, and D. W. Holt. Amiodarone disposition: Polyexponential, power and gamma functions. *Eur. J. Clin. Pharmacol.* **26**:655-856 (1984).
16. V. K. Piotrovskii. Pharmacokinetic stochastic model with Weibull-distributed residence times of drug molecules in the body. *Eur. J. Clin. Pharmacol.* **32**:515-523 (1987).
17. S. Riegelman and P. Collier. The application of statistical moment theory to the evaluation of in vivo dissolution time and absorption time. *J. Pharmacokin. Biopharm.* **8**:509-534 (1980).
18. J. M. van Rossum and C. A. M. van Ginneken. Pharmacokinetic system dynamics. In E. Gladtke and H. Heimann (eds.), *Pharmacokinetics*, Fischer, Stuttgart, 1980, pp. 53-73.
19. B. Jorgensen. Statistical properties of the generalized inverse Gaussian distribution. *Lecture Notes in Statistics*, Vol. 9, Springer-Verlag, New York, 1982.
20. M. S. Roberts, J. D. Donaldson, and M. Rowland. Models of hepatic elimination: Comparison of stochastic models to describe residence time distributions and to predict the influence of drug distribution, enzyme heterogeneity, and systemic recycling of hepatic elimination. *J. Pharmacokin. Biopharm.* **16**:41-83 (1988).
21. D. R. Cox. *Renewal Theory*, Methuen, London, 1962.
22. M. Weiss. Theorems on log-convex disposition curves in drug and tracer kinetics. *J. Theor. Biol.* **116**:355-368 (1985).
23. M. Brown. Bounds, inequalities, and monotonicity properties for some specialized renewal processes. *Ann. Probab.* **8**:227-240 (1980).
24. M. Weiss and K. S. Pang. The dynamics of drug distribution as assessed by the second and third curve moments. *Eur. J. Pharmacol.* **183**:611-622 (1990).
25. M. Brown. Further monotonicity properties for specialized renewal processes. *Ann. Probab.* **9**:891-895 (1981).
26. M. Weiss. A note on the role of generalized inverse Gaussian distributions of circulatory transit times in pharmacokinetics. *J. Math. Biol.* **18**:95-102 (1984).

27. M. Weiss. Moments of physiological transit time distributions and the time course of drug disposition in the body. *J. Math. Biol.* **15**:305-318 (1982).

28. M. Weiss. Washout time versus mean residence time. *Pharmazie* **43**:126-127 (1988).

29. M. Weiss. Model-independent assessment of accumulation kinetics based on moments of drug disposition curves. *Eur. J. Clin. Pharmacol.* **27**:355-359 (1984).

30. J. M. van Rossum, J. E. G. M. de Bie, G. van Lingen, and H. W. A. Teeuwen. Pharmacokinetics from a dynamical systems point of view. *J. Pharmacokin. Biopharm.* **17**:365-392 (1989).

PHARMACOKINETIC PARAMETER ESTIMATION WITH STOCHASTIC DYNAMIC MODELS

David Z. D'Argenio and **Ruomei Zhang**

Department of Biomedical Engineering
University of Southern California

ABSTRACT

The pharmacokinetic parameter estimation problem is reexamined within the framework of stochastic dynamic systems. Using this formalism, two sources of uncertainty are incorporated into the parameter estimation procedure: measurement error and process or model error. Consideration is given to linear dynamic models, with both model and measurement error terms modeled as Gaussian random processes. The maximum likelihood estimate of the parameters is obtained by using the Kalman filter formulation of the model to compute the likelihood function which is then maximized by direct nonlinear optimization. This approach to maximum likelihood estimation, given process or model error as well as output error, is evaluated using several simulated pharmacokinetic parameter estimation problems.

INTRODUCTION

One of the more significant applications of mathematical models in pharmacokinetics and pharmacodynamics involves their use in estimating those properties of the kinetic and/or dynamic processes that are not amenable to direct measurement. These applications result, for the most part, in nonlinear parameter estimation problems that are generally solved using traditional statistical methods of nonlinear regression (e.g., least squares, maximum likelihood and Bayesian estimation). While important work has been reported recently on methods of robust parameter estimation (e.g., [1–8]), the techniques developed have also been based on a regression model formulation of the estimation problem. A limitation of this regression framework is that it allows for only a single source of random uncertainty, appearing in the regression model as an additive error term (directly or by transformation).

In this chapter the pharmacokinetic model is expressed as a stochastic dynamic system, consisting of sets of differential equations and algebraic output equations. Using this formalism, random error terms are included not only in the output equations (representing measurement error), but also in the differential equations (representing process or model error). We consider the case of linear dynamic models and develop the maximum likelihood estimator for unknown model parameters given both model and output error terms. Several simulated pharmacokinetic parameter estimation problems are used to illustrate the method and to compare its results to those of a traditional regression model formulation for the estimation problem.

Advanced Methods of Pharmacokinetic and Pharmacodynamic Systems Analysis
Edited by D'Argenio, Plenum Press, New York, 1991

STOCHASTIC DYNAMIC MODEL FORMULATION

Consider first the case of deterministic linear, time-invariant pharmacokinetic models written in differential equation form as follows:

$$\dot{x}(t) = A(\phi)x(t) + B(\phi)r(t), \qquad x(0) = c \tag{1}$$

In this equation, $x(t)$ (dim n) is the state vector (e.g., compartment amounts, tissue concentrations), $r(t)$ (dim k) is the input vector (e.g., infusion rates), ϕ (dim p) is the vector of unknown model parameters (assumed to be time-invariant), $A(\phi)$ (dim $n \times n$) and $B(\phi)$ dim ($n \times k$) are the matrices defining the state and input connections, t is time, and $\dot{x}(t)$ represents the time derivative of the state vector ($dx(t)/dt$). The following output equation defines those functions of the states that are observable:

$$y(t) = C(\phi)x(t) \tag{2}$$

where $y(t)$ (dim l) is the output vector which is related to the states through the matrix $C(\phi)$ (dim $l \times n$). The system measurements, which occur at discrete times, $t_j, j = 1, \ldots, m$, are related to the model outputs as follows:

$$z(t_j) = y(t_j) + v(t_j), \qquad j = 1, \ldots, m \tag{3}$$

where $v(t_j)$ represents error. The problem then is to estimate the unknown model parameters given the system observations and the model defined by Eqs. (1) and (2). (It is assumed that the model is such that ϕ is identifiable.)

This dynamic model framework for the estimation problem has an equivalent nonlinear regression model formulation

$$z(t_j) = f(\phi, t_j) + v(t_j), \qquad j = 1, \ldots, m \tag{4}$$

where the function f is the solution (analytical or numerical) of Eqs. (1) and (2). Equation (4) is the traditional formulation for the individual estimation problem in pharmacokinetics for which several statistical estimation procedures are available. It should be emphasized that regression model procedures require that the model structure ($f(\phi, t)$ or Eqs. (1) and (2)) describes the system under study exactly and that any deviation between the model output $y(t)$ and its measured value $z(t)$ is due to the additive error $v(t)$.

As a generalization of Eq. (1), we consider the case of stochastic, linear time-invariant dynamic models as represented using the following notation:

$$\dot{x}(t) = A(\phi)x(t) + B(\phi)r(t) + w(t) \tag{5}$$

where $w(t)$ is a zero-mean Gaussian white noise process. (For the more rigorous stochastic differential equation formulation of Eq. (5) see the chapter by Schumitzky in this volume and [9].) The same model output and measurement equations given above (Eqs. (2) and (3)) are used here and can be combined as follows:

$$z(t_j) = C(\phi)x(t_j) + v(t_j) \tag{6}$$

Equations (5) and (6) define the linear stochastic model to be used as the framework for the pharmacokinetic estimation problem. In Eq. (6) the random variable $v(t_j)$ represents errors associated with the measurement procedure. The random process $w(t)$ appearing in Eq. (5) can be interpreted in two ways. It may represent some underlying physiological random variability that affects the evolution of the state of the system (process noise). Alternatively, the random vector $w(t)$ can be viewed as representing unmodeled dynamics, attempting to reflect the fact that the dynamic model in Eq. (1) does not describe the pharmacokinetic process exactly (model error). It is assumed, however, that the dynamic model does reflect the process under study with sufficient fidelity to allow the model parameters to adequately represent the kinetic properties we are interested in estimating.

MAXIMUM LIKELIHOOD ESTIMATION WITH MODEL AND OUTPUT ERROR

The estimation procedure will be developed using the following discrete-time formulation for the linear stochastic model of Eqs. (5) and (6):

$$x_{k+1} = A_k(\phi)x_k + B_k(\phi)r_k + w_k, \qquad x_0 = c \tag{7}$$

$$z_k = y_k + v_k = C(\phi)x_k + v_k \tag{8}$$

where the state, input, measurement and parameter vectors (x, r, z and ϕ), as well as the output matrix $C(\phi)$, are as defined above. In what follows the initial condition is assumed known, although this assumption can be removed. The matrices can be readily obtained from the continuous-time state and input matrices when $r(t)$ is piece-wise constant. See [9], for example, for a general derivation of an equivalent discrete-time stochastic model corresponding to Eq. (5). The chapter by Schumitzky in this volume illustrates this equivalence for a specific pharmacokinetic model.

In the above model, the sequences $\{w_k\}$ and $\{v_k\}$ are each assumed to be independent, zero-mean, Gaussian white-noise sequences with covariances Q_k and R_k, respectively; it is further assumed that $\{w_k\}$ and $\{v_k\}$ are mutually independent. (We have chosen not to relate the random sequence $\{w_k\}$ to the stochastic process $w(t)$ of the continuous model, although as mentioned there is an equivalent discrete-time representation for the continuous-time model in Eqs. (5) and (6).) As written, the discrete model in Eqs. (7) and (8) implies that the observation times and dose times occur simultaneously. While this restriction can be removed (as is done in the examples below), the following development will proceed using Eqs. (7) and (8) so as not to introduce distracting notation.

To calculate the maximum likelihood estimate $\widehat{\phi}$ for the parameters ϕ, given the data $z_1, \ldots, z_m$, it is necessary to compute the likelihood function of ϕ, $l(\phi|z_1, \ldots, z_m)$ (that is the conditional joint density function $p(z_1, z_2, \ldots, z_m|\phi)$. Due to the presence of the process noise, however, the measurements are correlated; that is, z_k depends on $w_{k-1}, w_{k-2}, \ldots$, just as it depends on past values of the dosage regimen. Therefore, $p(z_1, z_2, \ldots, z_m|\phi) \neq p(z_1|\phi)p(z_2|\phi)\cdots p(z_m|\phi)$, as would be the case if only the output error v_k were present.

A solution to this problem of computing the conditional joint density arises by considering the following Kalman filter formulation of the model defined in Eqs. (7) and (8):

$$\begin{aligned} \widehat{x}_{k+1} &= [A_k(\phi) - K_k(\phi)C(\phi)]\widehat{x}_k + B_k(\phi)r_k + K_k z_k, \qquad \widehat{x}_0 = x_0 &(9) \\ e_k &= -\widehat{y}_k + z_n = -C(\phi)\widehat{x}_k + z_k &(10) \end{aligned}$$

where e_k, $k = 1, \ldots, m$ is a sequence of mutually uncorrelated, zero-mean Gaussian random vectors with covariance

$$S_k(\phi) = C(\phi)P_k(\phi)C^T(\phi) + R_k(\phi) \tag{11}$$

The matrices $K_k(\phi)$ (dim $n \times l$) and $P_k(\phi)$ (dim $n \times n$) are defined as follows:

$$\begin{aligned} K_k(\phi) &= A_k(\phi)P_k(\phi)C^T(\phi)\left[C(\phi)P_k(\phi)C^T(\phi) + R_k\right]^{-1} &(12) \\ P_{k+1}(\phi) &= A_k(\phi)P_k(\phi)A^T(\phi) + Q_k \\ &\quad -K_k(\phi)\left[C(\phi)P_k(\phi)C^T(\phi) + R_k\right]K_k^T(\phi), \qquad P_0 = 0 &(13) \end{aligned}$$

The joint density of the sequence $\{e_k\}$ can be written

$$p(e_1, \ldots, e_m|\phi) = (2\pi)^{-lm/2} \prod_{k=1}^{m} [\det S_k(\phi)]^{-1/2} e^{-1/2[z_k - C(\phi)\widehat{x}_k]^T S_k^{-1}(\phi)[z_k - C(\phi)\widehat{x}_k]} \tag{14}$$

Numerous references are available that present the Kalman filter from an engineering systems theory viewpoint (e.g., [9–14]). For a discussion of the Kalman filter from a mathematical statistics perspective see [15–19].

Inspection of Eqs. (9) and (10) (also Eqs. (11)–(13)) reveals that for a given ϕ, e_k can be obtained from the measurements $z_k, z_{k-1}, \ldots, z_1$ and the original model (Eqs. (7) and (8)). Furthermore, Eqs. (9) and (10) can be rearranged to show that the sequence of measurements $\{z_k\}$ can be obtained given the sequence $\{e_k\}$ (for given ϕ). Thus, one can calculate $\{e_k\}$ given $\{z_k\}$ and inversely, $\{z_k\}$ given $\{e_k\}$; these two random sequences are therefore said to be causally invertible. Accordingly, the conditional joint density in Eq. (14) must contain the same information as $p(z_1, z_2, \ldots, z_n|\phi)$, and the likelihood function of the parameters given the measurements can be replaced by $l(\phi|e_1, \ldots, e_m)$.

The maximum likelihood estimate $\widehat{\phi}$ for the parameters of the model, given by Eqs. (7) and (8), can therefore be defined as follows:

$$\widehat{\phi} = \arg\left\{\max_{\phi \in \Phi} L(\phi|z_1, \ldots, z_m)\right\} \tag{15}$$

where

$$L(\phi|z_1, \ldots, z_m) = -\frac{1}{2}\sum_{k=1}^{m} \ln\det S_k(\phi) - \frac{1}{2}\sum_{k=1}^{m} [z_k - C(\phi)\widehat{x}_k]^T S_k^{-1}(\phi)[z_k - C(\phi)\widehat{x}_k] \tag{16}$$

This latter equation is simply the natural logarithm of Eq. (14) up to an additive constant. Of course, evaluation of Eq. (16) for a given ϕ requires Eqs. (9)–(13). Also, in defining the maximum likelihood estimate in Eq. (15) it is assumed that Q_k

and R_k, $k = 1, \ldots, m$, are known. A number of nonlinear optimization approaches can be used to solve the maximization problem (or an equivalent minimization problem) posed in Eq. (15). In the examples presented below, the Nelder-Mead simplex algorithm is used to solve Eq. (15). Given the maximum likelihood estimate $\widehat{\phi}$ defined in Eq. (15), the Kalman filter (Eqs. (9)–(13)) can then be used to estimate the mean ($\widehat{y}_k$, Eq. (10)) of the model output (y_k, Eq. (8)) given the observations.

To illustrate the proposed dynamic system maximum likelihood estimation procedure, given both process and output noise, several simulated pharmacokinetic parameter estimation problems will be considered. The first involves a one-compartment model representing the pharmacokinetics of intravenously administered theophylline. In the second example, a two-compartment model for the kinetics of teniposide is simulated. The third example involves a simulation of a continuous intravenous lidocaine infusion regimen. In these examples the simulated data are generated given both output and process error.

THEOPHYLLINE EXAMPLE

This first example is used largely to illustrate the influence of process error on measured plasma drug concentrations in relation to output or measurement error. We also use this example to contrast the behavior of the maximum likelihood estimation procedure outlined above to that of a regression model estimator (i.e., one which ignores the process error).

The one-compartment, two parameter model used to represent theophylline's pharmacokinetics is shown in Fig. 1, where $w(t)$ denotes the process noise input. The infusion input, represented by $r(t)$, is also defined in the figure along with the values for the kinetic parameters used in the simulation from [20]. The equations describing the stochastic discrete-time model corresponding to the deterministic compartment model shown in Fig. 1 are as follows:

$$x_{k+1} = e^{-CL(t_{k+1}-t_k)/V}\, x_k + \int_{t_k}^{t_{k+1}} e^{-CL(\tau - t_k)/V}\, r(\tau)d\tau + w_k, \qquad x_0 = 0 \tag{17}$$

$$z_k = x_k/V + v_k \tag{18}$$

where x represents drug amount and z measured drug concentration. Four observation times are considered, with $t_1 = 1.5, t_2 = 3.0, t_3 = 6.0$ and $t_4 = 12.0$ hrs. The zero-mean Gaussian process noise and output noise terms are assumed to be time-invariant with variances equal to 60^2 and 0.5^2, respectively.

Figures 2 and 3 illustrate the results of simulating the model defined by Eqs. (17) and (18). In Fig. 2, the mean (x) and standard deviation (error bars) of the drug concentrations are shown at each of the four observation times, as obtained from 1000 realizations of the random sequences. The output of the deterministic model is shown as the solid curve. Since the standard deviation of the measurement error is only 0.5, the process error contributes significantly to the variability of the observations (given the value chosen for the process error variance). As discussed above, the presence of the dynamic error process w_k also causes the observations to be correlated. The degree of intersample correlation is illustrated in Fig. 3, as calculated from the 1000 simulated data sets. The correlation between z_1 and each of three subsequent observations is shown by the solid squares on the lower curve. Correlations between z_2 and z_3, z_2 and z_4, and between z_3 and z_4 are also shown. Inspection of Fig. 3 indicates

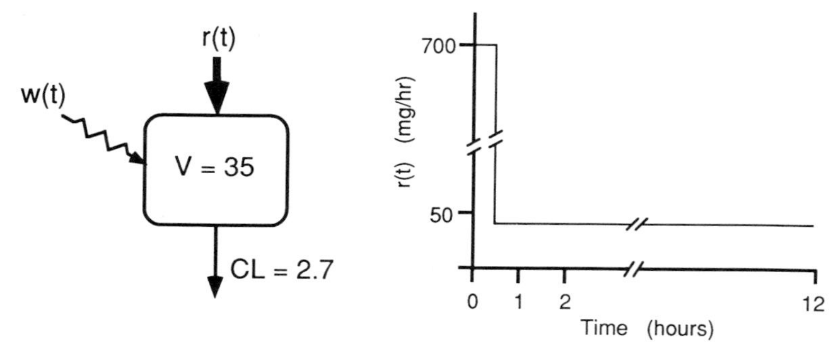

Fig. 1 Compartment model and infusion regimen for the theophylline example. V in L and CL in L/hr.

that the magnitude of the intersample correlation does not only depend on the time interval between the observations, but is also a function of the model dynamics in that particular time interval. For example, the observations at t_3 and t_4 are more highly correlated (0.64) than the observations at t_1 and t_3 (0.46) even though $t_4 - t_3$ is 6 hrs while $t_3 - t_1$ is 4.5 hrs.

To illustrate the influence that process error has on parameter estimation, Eqs. (17) and (18) were simulated with $E(v_k^2) = 0$ (no observation error) and $E(w_k^2) = 60^2$. Figure 4 shows a sample data set, with the exact simulated concentrations indicated by the symbols (■) (again using $CL = 2.7$ L/hr and $V = 35$ L). The corresponding maximum likelihood parameter estimates (using the dynamic system framework presented above) are $\widehat{CL} = 2.2$ L/hr and $\hat{V} = 33.2$ L, with the resulting predicted model output shown in Fig. 4 by the symbols $\triangle$. For comparison, the least squares parameter estimates were also calculated from these data by adopting a regression model formulation for the estimation problem (i.e., assuming all the uncertainty is

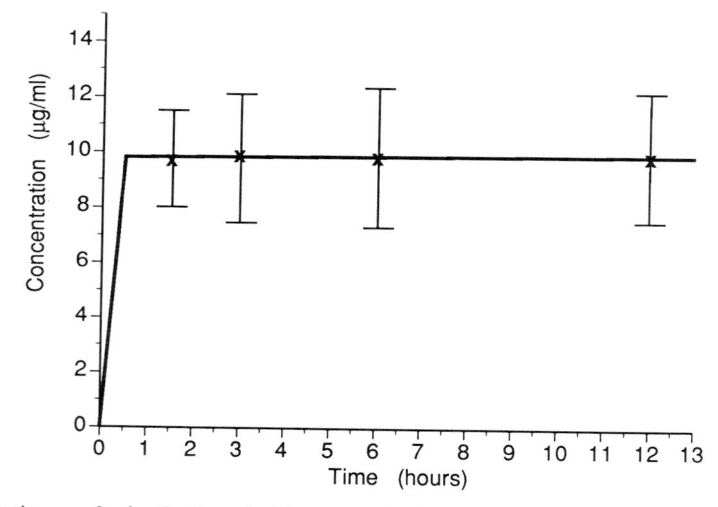

Fig. 2. Simulation results for the theophylline example showing the output of the deterministic model for the parameter values and input given in Fig. 1 (solid line). Also shown at each of the four observation times is the mean (x) and standard derivation (error bar) obtained from the data sets simulated from Eqs. (17) and (18).

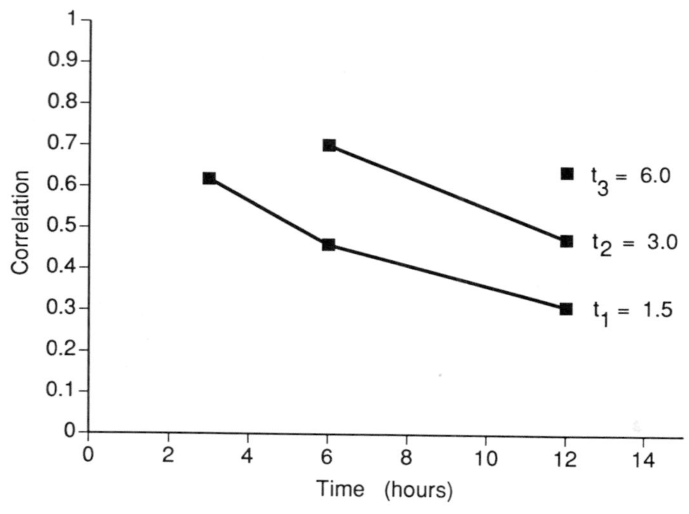

Fig. 3 Inter-observation correlation for the theophylline example as calculated from the 1000 simulated data sets.

due to output error – of course, in this case there is no output error and all the uncertainty results from the process noise). The parameter estimates obtained using the regression model estimator are $\widehat{CL} = 1.7$ L/hr and $\widehat{V} = 29.3$ L with the model predictions shown by the symbol ◇ in Fig. 4.

As a second example, Eqs. (17) and (18) were used to simulated plasma concentration measurements with both process noise and output error present (same model parameters and statistics as used to construct Figs. 2 and 3). The resulting simulated observations are shown in Fig. 5 by the symbol ●, while the exact concentrations (i.e., before adding output error) are indicated by the symbol ■. Applying the maximum likelihood estimation procedure to the simulated observations yields parameters estimates $\widehat{CL} = 2.5$ L/hr and $\widehat{V} = 39.3$ L, with the corresponding predicted plasma

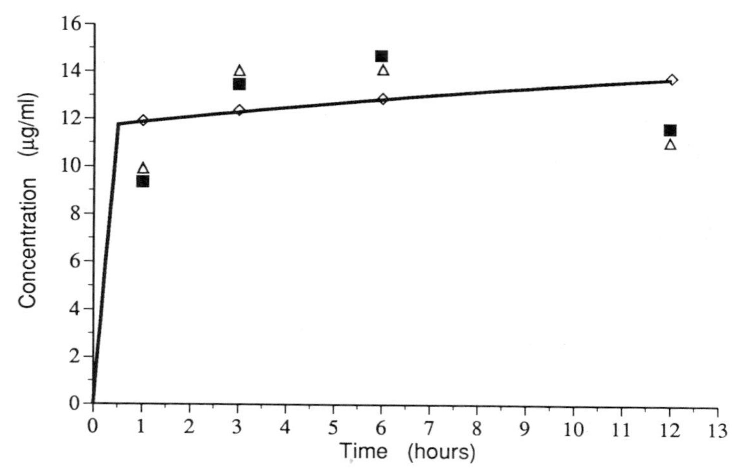

Fig. 4. Estimation results for the theophylline example with no observation error. ■ – exact model output; △ – maximum likelihood estimate of plasma concentration; ◇ – regression model estimate of plasma concentration (also continuous curve).

concentrations shown in Fig. 5 by the symbol $\triangle$. The least squares regression model approach results in parameter estimates $\widehat{CL} = 2.6$ L/hr and $\widehat{V} = 57.2$ L, which predict the concentrations shown in Fig. 5 by the symbol $\diamond$. This example illustrates that the dynamic system estimation procedure attributes some of the variability in the observations to measurement error, but also attributes some of the variability (a large portion in this example) to process noise.

TENIPOSIDE EXAMPLE

The second example involves a two-compartment model for the plasma kinetics of the anticancer drug teniposide as shown in Fig. 6. Also given in the figure is the drug infusion regimen as well as values for the kinetic parameters (from [21]). The following equations describe the stochastic discrete-time model corresponding to the compartment model shown in Fig. 6:

$$x_{k+1} = \Phi(t_{k+1} - t_k)x_k + \int_{t_k}^{t_{k+1}} \Phi(\tau - t_k) \begin{bmatrix} r(\tau) \\ 0 \end{bmatrix} d\tau + w_k \quad , x_0 = 0 \qquad (19)$$

$$z_k = y_k + v_k = [1/V \;\; 0]\, x_k + v_k \qquad (20)$$

where the vector x_k (dim 2) represents the amount of drug in the central (first entry) and peripheral (second entry) compartments at time t_k. The matrix Φ (dim 2×2) is given as follows:

$$\Phi(t) = \frac{1}{\lambda_1 - \lambda_2} \cdot$$

$$\begin{bmatrix} (\lambda_1 - K_{pc})e^{-\lambda_1 t} + (K_{pc} - \lambda_2)e^{-\lambda_2 t} & -K_{pc}e^{-\lambda_1 t} + K_{pc}e^{-\lambda_2 t} \\ -K_{cp}e^{-\lambda_1 t} + K_{cp}e^{-\lambda_2 t} & (\lambda_1 - K_e - K_{cp})e^{-\lambda_1 t} + (K_e + K_{cp} - \lambda_2)e^{-\lambda_2 t} \end{bmatrix} \qquad (21)$$

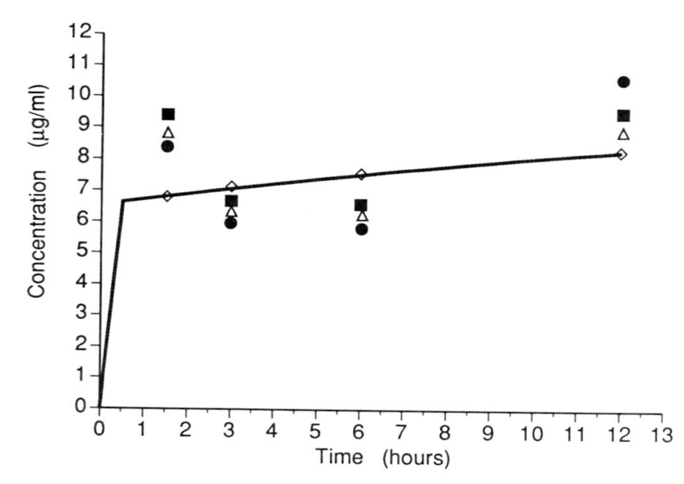

Fig. 5. Estimation results for the theophylline example with both process noise and output error. ●– simulated measured plasma concentrations. See Fig. 4 caption for definition of other symbols.

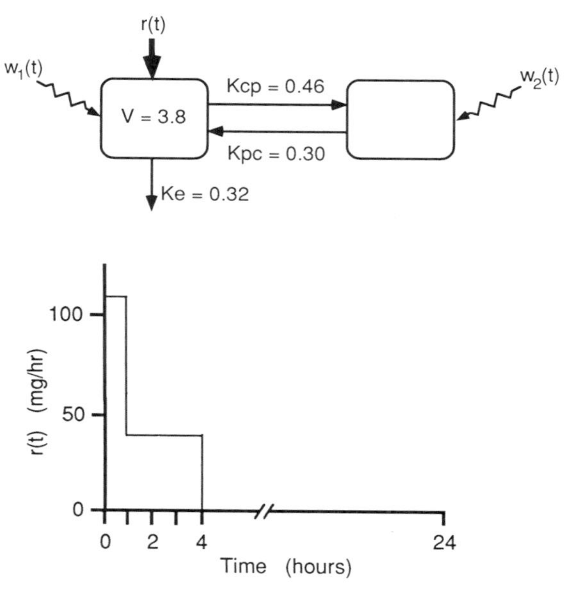

Fig. 6. Compartment model and infusion regimen for the teniposide example. V in L and rate constants in hr^{-1}.

where $\lambda_1 = (K + \sqrt{K^2 - 4K_eK_{pc}})/2$, $\lambda_2 = (K - \sqrt{K^2 - 4K_eK_{pc}})/2$ and $K = K_e + K_{cp} + K_{pc}$. The process noise and output noise were assumed to be time-invariant (i.e., $Q_k = Q$ (dim 2×2) and $R_K = R$).

The output of the model defined by Eqs. (19) and (20) was simulated at 10 observation times on the interval 1.0 to 20.0 hrs (1.0, 2.5, 4.0, 4.5, 5.0, 6.0, 8.0, 12.0, 16.0, 20.0). Figure 7 shows the mean (x) and standard deviation (error bars) of the teniposide plasma concentrations obtained from 1000 realizations for the process and output noise sequences with $Q = \text{diag}\{9 \quad 36\}$ and $R = 1$. The continuous curve shows the response of the deterministic model given the parameter values and infusion input from Fig. 6. The contribution of the process noise to the variability of the simulated plasma concentrations depicted in Fig. 7 ranges from 13% to 40%. Table I gives the intersample correlation for this example.

For each of the 1000 simulated data sets the maximum likelihood estimates of the model's parameters were calculated as described previously. The least squares regression model parameter estimates were also calculated for each of the simulated data sets. The first two rows in Table II list the resulting means and standard derivations for the four model parameters from both of these estimation procedures. These results indicate that the dynamic system maximum likelihood estimates are less biased and significantly less variable than the regression model estimates obtained by ignoring the contribution of process error.

Additional data sets were simulated from Eqs. (19) and (20) (1000 realizations each) for three other process error covariances: $Q = \text{diag}\{4\ 16\}$, $Q = \text{diag}\{1\ 4\}$, and $Q = \text{diag}\{0\ 0\}$ (i.e., no process uncertainty). For all cases the output error variance was $R = 1.0$. In the first of these cases, the process noise contribution to the standard

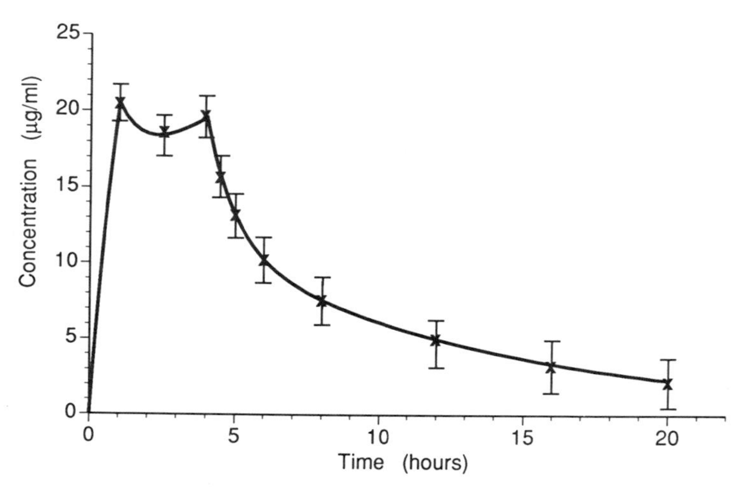

Fig. 7 Simulation results for the teniposide example showing output of the deterministic system (solid line) and mean (**x**) and standard derivation (error bars) for the simulated measurements obtained with $Q = \text{diag}\{9\ 36\}$ and $R = 1$.

Table I. Inter-Observation Correlation for the Teniposide Example Calculated from the 1000 Simulated Data Sets ($Q = \text{diag}\{9\quad 36\}$, $R = 1$). See Text for Observation Times $t_1, \ldots, t_{10}$.

	t_1	t_2	t_3	t_4	t_5	t_6	t_7	t_8	t_9
t_2	0.15								
t_3	0.11	0.25							
t_4	0.07	0.19	0.42						
t_5	0.08	0.16	0.38	0.42					
t_6	0.13	0.22	0.31	0.38	0.38				
t_7	0.11	0.17	0.25	0.29	0.32	0.38			
t_8	0.01	0.14	0.17	0.22	0.20	0.28	0.34		
t_9	0.05	0.09	0.10	0.16	0.16	0.16	0.26	0.35	
t_{10}	0.00	0.03	0.07	0.07	0.09	0.07	0.14	0.22	0.38

Table II. Maximum Likelihood (ML) and Regression Model (RM) Estimation Results for the Teniposide Example. True parameter values are $K_e = 0.32\ \text{hr}^{-1}$, $V = 3.8$ L, $K_{cp} = 0.46\ \text{hr}^{-1}$ and $K_{pc} = 0.30\ \text{hr}^{-1}$.

Process Error		Estimation Results			
Variance	Estimator	$\widehat{K_e}(\text{hr}^{-1})$	$\widehat{V}(\text{L})$	$\widehat{K_{cp}}(\text{hr}^{-1})$	$\widehat{K_{pc}}(\text{hr}^{-1})$
$Q = \text{diag}\{9\ 36\}$	RM	0.37±0.34	3.7±0.70	0.85±1.9	0.36±0.25
	ML	0.31±0.14	3.8±0.58	0.58±0.57	0.30±0.20
$Q = \text{diag}\{4\ 16\}$	RM	0.35±0.22	3.7±0.58	0.67±1.3	0.34±0.19
	ML	0.32±0.11	3.8±0.53	0.56±0.48	0.31±0.17
$Q = \text{diag}\{1\ 4\}$	RM	0.33±0.13	3.7±0.49	0.57±0.74	0.32±0.15
	ML	0.32±0.09	3.8±0.47	0.54±0.37	0.32±0.14
$Q = \text{diag}\{0\ 0\}$	RM,ML	0.32±0.07	3.8±0.43	0.52±0.23	0.32±0.13

deviation of the simulated plasma concentrations ranged from 5 to 26%, while in the second case the process noise contribution was between 1 and 11%. The dynamic system maximum likelihood (ML) and regression model (RM) parameter estimation procedures were also applied to these data sets, with the results summarized in Table II. These results illustrate that, as expected, the difference between the maximum likelihood and regression model parameter estimates decreases as the process noise covariance decreases. The results summarized in Table II show that incorporating process noise in the simulated data has a significant influence on the precision of the estimated parameters regardless of which estimation procedure is used. This is even the case for the lowest (nonzero) process covariance considered. It is also evident from Table II that for all process error covariances considered the dynamic system maximum likelihood estimation procedure results in estimates that are less biased and variable than those obtained using the regression model estimator.

LIDOCAINE EXAMPLE

In this example the two-compartment model shown in Fig. 6 is used to simulate the plasma kinetics of lidocaine in response to a continuous intravenous infusion. The dose regimen $r(t)$ for this example is given as follows: $r(t) = 75$ mg/min for $0.0 < t < 1.0$ min and $r(t) = 1.45$ mg/min for $t > 1.0$ min. Values for the model's parameters were taken from a study of lidocaine kinetics in normal subjects [22]: $K_e = 0.0242\ \text{min}^{-1}$, $K_{cp} = 0.066\ \text{min}^{-1}$, $K_{pc} = 0.038\ \text{min}^{-1}$, $V = 30.0$ L. The dose regimen is designed to achieve and maintain a plasma concentration of 2.0 μg/ml. This model was used to simulate plasma concentrations at the following 10 observation times during the maintenance infusion: 2.0, 5.0, 10.0, 30.0, 60.0, 90.0, 120.0, 180.0, 360.0, 720.0 min.

Table III shows the mean and standard deviation of the plasma concentrations obtained from 1000 simulated data sets with $Q = \text{diag}\{9\ 4\}$ and $R = 0.04$. As shown in Table III, the process error is responsible for between 11 to 52% of the variability present in the simulated lidocaine plasma concentration data. The inter-observation correlations for this example are given in Table IV. Data sets for three additional process noise covariances were also simulated: $Q = \text{diag}\{4\ 2\}$, $Q = \text{diag}\{2\ 1\}$, and $Q = \text{diag}\{0\ 0\}$ ($R = 0.04$ and 1000 realizations in each case). In the first of these last three cases the process noise contributed between 4 and 37% to the standard deviation of the simulated lidocaine concentrations, and between 0.5 and 25% in the second case. The last case, in which no process error was included in the simulated data, again provides a basis for assessing the effect of process noise on the estimation procedures considered.

The estimation results obtained using these data sets are summarized in Table V for both the dynamic system maximum likelihood (ML) and regression model (RM) parameter estimation procedures. These results are similar in most respects to those obtained in the teniposide example. In this example the addition of process noise also has an important influence on the variability of the parameter estimates for both estimation procedures. Futhermore, the dynamic system maximum likelihood estimates are again more precise than the regression model estimates for the two cases with the greatest process error covariance. There is little difference, however, between the methods for the case with $Q = \text{diag}\{2\ 1\}$, with both methods resulting in substantial estimator variability compared to the case of no process noise.

Table III. Simulated Plasma Concentrations (Mean ± Standard Deviations of 1000 Data Sets) for the Lidocaine Example with $Q = \text{diag}\{9\ \ 4\}$ and $R = 0.04$.

Time (min)	Concentration (μg/ml)	Time (min)	Concentration (μg/ml)
2	2.2±0.22	90	1.5±0.39
5	1.9±0.28	120	1.6±0.39
10	1.5±0.30	180	1.7±0.42
30	1.2±0.34	360	1.9±0.42
60	1.3±0.36	720	2.0±0.41

Table IV. Inter-Observation Correlation for the Lidocaine Example ($Q = \text{diag}\{9\ \ 4\}$, $R = 0.04$). See Text for Observation Times $t_1, \ldots, t_{10}$.

	t_1	t_2	t_3	t_4	t_5	t_6	t_7	t_8	t_9
t_2	0.22								
t_3	0.15	0.29							
t_4	0.04	0.15	0.19						
t_5	0.02	0.09	0.13	0.31					
t_6	0.02	0.04	0.04	0.19	0.28				
t_7	0.03	0.03	0.08	0.12	0.23	0.39			
t_8	0.04	0.06	0.07	0.07	0.15	0.20	0.30		
t_9	0.02	0.08	0.03	0.02	0.01	0.03	0.07	0.08	
t_{10}	0.02	0.02	0.02	0.01	0.02	0.01	0.01	0.05	0.09

Table V. Maximum Likelihood (ML) and Regression Model (RM) Estimation Results for the Lidocaine Example. True parameter values are $K_e = 0.0242\ \text{min}^{-1}$, $V = 30$ L, $K_{cp} = 0.066\ \text{min}^{-1}$ and $K_{pc} = 0.038\ \text{min}^{-1}$.

Process Error Variance	Estimator	Estimation Results			
		$\widehat{K_e}(\text{min}^{-1})$	$\widehat{V}$(L)	$\widehat{K_{cp}}(\text{min}^{-1})$	$\widehat{K_{pc}}(\text{min}^{-1})$
$Q = \text{diag}\{9\ 4\}$	RM	0.029±0.047	30.0±5.6	0.13±0.27	0.059±0.067
	ML	0.024±0.0093	30.0±4.7	0.10±0.11	0.056±0.070
$Q = \text{diag}\{4\ 2\}$	RM	0.026±0.025	30.0±4.7	0.097±0.17	0.050±0.049
	ML	0.024±0.0070	30.0±4.2	0.087±0.086	0.049±0.054
$Q = \text{diag}\{2\ 1\}$	RM	0.025±0.0064	30.0±4.1	0.082±0.081	0.046±0.038
	ML	0.025±0.0062	30.0±4.1	0.083±0.082	0.047±0.045
$Q = \text{diag}\{0\ 0\}$	RM,ML	0.024±0.0045	30.0±3.7	0.074±0.040	0.041±0.025

DISCUSSION

An approach to pharmacokinetic model parameter estimation that formally incorporates both measurement error uncertainty as well as process or model uncertainty has been proposed and illustrated. Model uncertainty is represented by a random process added to the differential equations defining the pharmacokinetic model. The case of maximum likelihood estimation involving linear dynamic models is considered, with both random error terms modeled as zero mean, Gaussian processes. Given a set of measurements, the Kalman filter representation of the kinetic model is used to construct the likelihood function of the unknown parameters which is then maximized using a nonlinear optimization procedure.

In addition to demonstrating the feasibility of implementing the proposed dynamic system maximum likelihood estimation procedure, the simulations presented serve to illustrate several important consequences of including process noise in pharmacokinetic models. First, as exemplified in the theophylline model example, the presence of process noise causes the observations to be correlated (see Fig. 3), even though the process and output errors are themselves independent. Second, even a small amount of process noise – as judged by its contribution to the standard derivation of the observations – can result in important differences in estimated parameters compared to the case without process error (see teniposide and lidocaine examples, Tables I and IV). Third, ignoring process noise as a source of variability in pharmacokinetic measurements can lead to parameter estimates with unnecessarily large variances (see Tables I and IV). Whether these observations are unique to the examples considered herein remains to be explored.

The use of the Kalman filter for parameter estimation involving linear stochastic dynamic models is not new in engineering systems theory. The approach generally considered involves including the unknown model parameters as state variables in a new augmented dynamic model (now nonlinear). The resulting state estimation problem is then handled by combining dynamic model linearization with Kalman filter theory. Many variations of this basic idea have been proposed and are often called *extended Kalman filters* (e.g., [23–26]). This approach has also been applied to the problem of pharmacokinetic model parameter estimation [28]. The dynamic system parameter estimation method proposed herein differs fundamentally from the extended Kalman filter approach. In contrast to the extended Kalman filter, the method introduced above does not involve a linearizing approximation; instead the Kalman filter is used only to form the nonlinear likelihood function for the unknown parameters, which is then extremized using a nonlinear optimization procedure.

A number of important extensions of the proposed dynamic system maximum likelihood parameter estimation procedure are necessary if the method is to be applicable to a broad class of pharmacokinetic estimation problems. The case of state dependent output error needs to be considered since the variance of many drug assay procedures is a function of drug concentration. Also, methods for analysis of estimator error associated with the dynamic system maximum likelihood estimation procedure should be examined. Other formulations of the Kalman filter, such as various smoothing implementations, should also be considered as they may yield improved small sample estimator characteristics. The stochastic dynamic system estimation framework can also be applied to the case of least squares and Bayesian (*maximum a posterior probability*) estimation. Finally, extension to the case of nonlinear dynamic systems, while difficult, requires investigation.

ACKNOWLEDGMENTS

The authors wish to thank Alan Schumitzky for many exciting discussions on the theory of stochastic dynamic systems. This work was supported in part by grant P41 RR01861 from the National Institutes of Health.

REFERENCES

1. L. Endrenyi. Diagnosis, robust design and estimation in nonlinear regression. *T. 42nd Annual Quality Control Conf.*, Rochester, 1986, pp. 21–41.
2. R. J. Carroll and D. Ruppert. A comparison between maximum likelihood and generalized least squares in a heteroscedastic linear model. *J. Am. Stat. Assoc.* **77**:878–882 (1982).
3. R. J. Carroll and D. Ruppert. Robust estimation in heteroscedastic linear models. *Ann. Stat.* **10**:429–441 (1984).
4. D. M. Giltinan, R. J. Carroll, and D. Ruppert. Some new estimation methods for weighted regression when there are possible outliers. *Technometrics* **28**:219–230 (1986).
5. R. J. Carroll and D. Ruppert. Diagnostics and robust estimation when transforming the regression model and the response. *Technometrics* **29**:287–299 (1987).
6. S. L. Beal and L. B. Sheiner. Heteroscedastic nonlinear regression. *Technometrics* **30**:327–338 (1988).
7. D. M. Giltinan and D. Ruppert. Fitting heteroscedastic regression models to individual pharmacokinetic data using standard statistical software. *J. Pharmacokin. Biopharm.* **17**:601–614 (1989).
8. D. M. Bates and D. G. Watts. *Nonlinear Regression Analysis and Its Application*, Wiley, New York, 1988.
9. P. S. Maybeck. *Stochastic Models, Estimation and Control*, Volume 1, Academic Press, New York, 1979.
10. P. S. Maybeck. *Stochastic Models, Estimation and Control*, Volume 2, Academic Press, New York, 1982.
11. F. S. Schweppe. *Uncertain Dynamic Systems*, Prentice Hall, Inc., Englewood Cliffs, 1973.
12. G. C. Goodwin and R. L. Payne. *Dynamic System Identification: Experiment Design and Data Analysis*, Academic Press, New York, 1977.
13. G. C. Goodwin and K. S. Sin. *Adaptive Filtering Prediction on Control*, Prentice-Hall, Inc., Englewood Cliffs, 1984.
14. J. M. Mendel. *Lessons in Digital Estimation Theory*, Prentice-Hall, Inc., Englewood Cliffs, 1987.
15. P. J. Harrison and C. F. Stevens. A Bayesian approach to short-term forecasting. *Oper. Res. Quart.* **22**:341–362 (1971).
16. P. J. Harrison and C. F. Stevens. Bayesian forecasting (with discussion). *J. Roy. Stat. Soc. B. Met.* **38**:205–247 (1976).
17. R. J. Meinhold and N. D. Singpurwalla. Understanding the Kalman Filter. *Am. Stat.* **37**:123–127 (1983).
18. L. R. Weill and P. N. DeLand. The Kalman Filter: An introduction to the mathematics of linear least mean square recursive estimation. *Int. J. Math. Educ. Sci. Technol.* **17**:347–366 (1986).
19. J. C. Spall. *Bayesian Analysis of Time Series and Dynamic Models*, Marcel Dekker, Inc., New York, 1988.
20. J. R. Powell, S. Vozeh, P. Hopewell, J. Costello, L. B. Sheiner, and S. Riegelman. Theophylline disposition in acutely ill hospitalized patients. The effect of smoking, heart failure, severe airway obstruction, and pneumonia. *Am. Rev. Respir. Dis.* **118**:229–238 (1978).
21. J. H. Rodman, M. Abromowitch, J. A. Sinkule, F. A. Hayes, G. K. Rivera, and W. E. Evans. Clinical pharmacodynamics of continuous infusion teniposide: Systemic exposure as a determinant of response in a phase I trial. *J. Clin. Oncol.* Vol. 5, No. 7, 1007–1014 (1987).
22. M. Rowland, P. D. Thomson, A. Guichard, and K. L. Melmon. Disposition kinetics of lidocaine in normal subjects. *Ann. N. Y. Acad. Sci.* **179**:383–398 (1971).
23. A. H. Jazwinski. *Stochastic Processes and Filtering Theory*, Academic Press, New York, 1970.

24. A. P. Sage and C. D. Wakefield. Maximum likelihood identification of time varying and random system parameters. *Int. J. Contr.*, Vol. 16, No. 1, 81–100 (1972).

25. P. Eykhoff. *System Identification: Parameter and State Estimation*, John Wiley & Sons, Chicester, 1974.

26. L. W. Nelson and E. Stear. The simultaneous on-line estimation of parameters and states in linear systems. *IEEE T. Automat. Contr.* **AC-21**:94-98 (1976).

27. L. Ljung. Asymptotic behaviour of the extended Kalman filter as parameter estimator for linear systems. *IEEE T. Automat. Contr.* **AC-24**:36–51 (1979).

28. W. F. Powers, P. H. Abbrecht, and D. G. Covell. Systems and microcomputer approach to anticoagulant therapy. *IEEE T. Bio-med. Eng.* **27**:520–523 (1980).

RELATIONSHIPS BETWEEN INTRA- OR INTERINDIVIDUAL VARIABILITY AND BIOLOGICAL COVARIATES: APPLICATION TO ZIDOVUDINE PHARMACOKINETICS

France Mentré and **Alain Mallet**

Méthodologie Informatique et Statistique en Médecine
SIM-INSERM U 194, Paris

ABSTRACT

Quantification of interindividual variability and determination of the covariates contributing to this variability have proved to improve decision-making procedures and individualization of dosage regimen. Several methods have been developed to estimate the probability distribution of the pharmacokinetic parameters from measurements obtained in a sample of individuals. All these methods assume that parameters and covariates remain stationary within an individual. However, ascertaining the stationarity of the parameters or quantifying intraindividual variability can be important when chronic administration is scheduled. We propose a population approach to study the stationarity of the pharmacokinetic parameters. This method is developed around the Non-Parametric Maximum Likelihood estimation method and is based on the comparison of the likelihood of two samples of data from the same individuals at two different times under several assumptions. Some ideas to study stationarity of both individual pharmacokinetic parameters and covariates are also given. The proposed method is illustrated on a very simple simulated example which reflects continuous infusion of a drug involving various assumptions about the nonstationarity of the parameters. The approach is then applied to population analysis of zidovudine kinetics data measured on the first and 35th day of therapy in 36 patients.

INTRODUCTION

Quantification of interindividual variability and determination of the covariates contributing to this variability have proved to improve decision-making procedures and individualization of dosage regimen [1]. The usual approach, known as population pharmacokinetics, is to assume that pharmacokinetics parameters are randomly distributed across individuals and to estimate the probability distribution of these parameters. Given a pharmacokinetic model of the drug under study and given specification of the measurement noise, several methods have been developed to estimate the probability distribution of the parameters from routinely collected data (e.g., blood concentrations) in a sample of individuals [2]. A further step is to incorporate in the analysis the biological covariates (e.g., body weight, serum creatinine,

Advanced Methods of Pharmacokinetic and Pharmacodynamic Systems Analysis
Edited by D'Argenio, Plenum Press, New York, 1991

age) that are potential variables of interindividual variability and to estimate the statistical relationships between the covariates and the pharmacokinetic parameters. This approach allows Bayesian estimation of the parameters of a new individual given only the covariates values and/or one or more measurements [3].

All these methods assume that parameters and covariates remain stationary within an individual. However, ascertaining the stationarity of the parameters or quantifying intraindividual variability can be important when chronic administration is scheduled. When the number of individual measurements is limited, for example when only routinely collected trough concentrations are available, classical estimation methods are not appropriate. We, therefore, have tried to develop a population approach to study the stationarity of the pharmacokinetic parameters. The method is developed around the Non-Parametric Maximum Likelihood (NPML) estimation procedure [4] which is reviewed in the next section. The proposed method, detailed below, is based on the comparison of the likelihood of two samples of data from the same individuals at two different times under several assumptions. Some ideas to study stationarity of both individual pharmacokinetic parameters and covariates are also given. The method is then illustrated on a very simple simulated example which reflects continuous infusion of a drug. Various assumptions about the nonstationarity of the parameters are defined.

The problems associated with nonstationarity of drug disposition processes were raised while estimating the pharmacokinetic characteristics of zidovudine (currently AZT) in 36 AIDS patients from measurements performed on the first day and on the 35th day of therapy. The proposed approach suggests important intra and interindividual variability for this drug. The data and the results are described in the fourth paragraph.

THE NPML ESTIMATION METHOD

This *distribution-free* method for estimating the distribution of pharmacokinetic parameters has already been described [4] and used for the description of the population characteristics of various drugs [5–6]. The main features are summarized here to clarify the notation.

Let ϕ be the p-dimensional vector of the pharmacokinetic parameters of the drug. Let N be the total number of individuals in the sample and let y_j be the vector of the n_j measurements performed in the individual j. When the structural model of the pharmacokinetic process is given and when the measurement noise model is specified, the likelihood $l_j(y_j, \phi)$ of the data y_j given ϕ is readily defined.

For example, let $f_j(\phi)$ denote the vector of the n_j model-predicted concentrations in individual j for the value ϕ of the pharmacokinetic parameters. If e_j, the measurement error, is additive, with a zero-mean Gaussian distribution and a variance-covariance matrix R_j, then $l_j(y_j, \phi)$ is given by:

$$l_j(y_j, \phi) = (2\pi det(R_j))^{-\frac{n_j}{2}} e^{-\frac{1}{2}(y_j - f_j(\phi))^T R_j^{-1}(y_j - f_j(\phi))} \tag{1}$$

Assume now that the vector of individual parameters is a random variable Φ with probability distribution F. Given F, the likelihood of the data y_j of individual j can be written:

$$L_j(F) = \int l_j(y_j, \phi)F(\phi)d\phi = E_F\left[(l_j(y_j, \phi)\right] \tag{2}$$

where E_F is the expectation with respect to Φ.

For N independent individuals, the likelihood of the whole set of data $\{y_1, ..., y_N\}$ given F is therefore:

$$L(F) = \prod_{j=1}^{N} L_j(F) = \prod_{j=1}^{N} E_F\left(l_j(y_j, \phi)\right) \tag{3}$$

Maximization of this likelihood provides an estimate $\widehat{F}$ of the probability distribution of the pharmacokinetic parameters Φ. It has been established that $\widehat{F}$ is discrete, involving at most N locations ϕ_k with corresponding frequencies α_k.

This method provides a complete description of F without any *a priori* assumption on F and allows easy derivation of, for instance, the first two moments of $\widehat{F}$ according to:

$$E_{\widehat{F}}(\Phi) = \sum_k \alpha_k \phi_k \tag{4}$$

$$var_{\widehat{F}} = \sum_k \alpha_k \phi_k \phi_k^T \tag{5}$$

Furthermore, consider the case where for every individual we have, in addition to the concentration measurements y_j, a set of covariates q_j. Assuming that the covariates are randomly distributed across the population, the same method provides an estimate of the joint distribution of the pharmacokinetic parameters and of the covariates. Thus, the probability distribution of Φ conditioned on any value q of the covariates can be computed. It is, therefore, possible to describe the statistical relationships between the pharmacokinetic parameters and the covariates without specifying a regression model (also called second stage model).

LIKELIHOOD COMPARISONS FOR STATIONARITY ASSESSMENT

Assume we now have two samples of measurements $(y_j^a,\ j = 1, \ldots, N)$ and $(y_j^b,\ j = 1, \ldots, N)$ on the same N individuals with similar experimental designs performed on two different days D_a and D_b. When there are not enough measurements within each individual to estimate accurately the individual's parameters ϕ_j^a on day D_a and ϕ_j^b on day D_b, it is difficult to assess whether the individual parameters are stationary. We, therefore, propose to compare the maximum likelihood of the total set of data $(y_j^a, y_j^b, j = 1, \ldots, N)$ under two different hypotheses:

$$H_0: \quad \phi_j^a = \phi_j^b = \phi_j \quad \text{for all individuals}$$
$$H_1: \quad \phi_j^a \neq \phi_j^b \quad \text{for some individuals}$$

Under H_0, the parameters are assumed to be stationary between the two days, the likelihood L_0 of the total set can be computed as a function of the probability distribution F of Φ by:

$$L_0(F) = \prod_{j=1}^{N} E_F\left(f_j(y_j^a, \phi) f_j(y_j^b, \phi)\right) \tag{6}$$

Under H_1, the parameters are no longer similar. Let us then define a new parameter vector

$$\phi^{ab} = (\phi^a, \phi^b)$$

of dimension $2p$. The likelihood L_1 of the total set can be computed as a function of the probability distribution F of Φ^{ab} by:

$$L_1(F) = \prod_{j=1}^{N} E_F \left(f_j(y_j^a, \phi^a) f_j(y_j^b, \phi^b) \right) \tag{7}$$

where ϕ^a and ϕ^b are, respectively, the first and the last p components of the vector ϕ^{ab}.

The NPML estimation method provides optimal likelihoods under the two assumptions. It can be shown that L_1 is always greater than L_0 and that under the null hypothesis $L_0 \approx L_1$. However, no statistical test has yet been defined to accept or reject H_0 in the case of a small difference between the two likelihoods.

If the parameters do not appear to be similar, one can define another null hypothesis: H_0': for each individual y_j^a and y_j^b are independent. The parameters ϕ_j^a and ϕ_j^b are different but drawn for the same distribution F of Φ, that is to say the data y_j^a and y_j^b are considered as obtained from two different individuals. The likelihood L_0' of the $2N$ samples $(y_j^a) \cup (y_j^b)$ can be written as:

$$L_0'(F) = \prod_{j=1}^{N} E_F \left(f_j(y_j^a, \phi) \right) \prod_{j=1}^{N} E_F \left(f_j(y_j^b, \phi) \right) \tag{8}$$

It can be shown that L_1 is always greater than L_0' and that under H_0', i.e., independency between day D_a and day D_b, $L_0' \approx L_1$. If neither H_0 and H_0' are valid, there is no relationship between L_0 and L_0'. The computation of these values, however, may give some indication about the most likely hypothesis.

It should be noted that the assumption of similarity of the experimental designs on the two days was made for sake of simplicity of the notation. Of course, the same methodology applies when they are different; in such a case, the individual likelihoods on each day are based on the performed design.

As a more general case, we assume some biological covariates are to be included in the analysis that are related on the first day to the pharmacokinetic parameters. The same approach to ascertain both stationarity of parameters and covariates can be used even if it is often quite easy to check whether the covariates are stationary or not. However, it is interesting to study whether both modifications are related: for instance, is an increase in clearance related to the increase in body weight because of health improvement?

Four different cases can be distinguished:

1. Both parameters and covariates are stationary between the two days: the joint and conditional distribution are similar on days D_a and D_b.

2. Parameters are stationary but not the covariates, i.e., changes of covariates do not imply modifications of the parameters. The relationship between the parameters and the covariates on day D_a is no longer valid on day D_b.

3. The parameters are not stationary but the covariates remain constant. One cannot predict the parameters on day D_b from the joint distribution on day D_a.

4. Both parameters and covariates are not stationary. Again, one can distinguish two separate cases:

   a) there is no relationship between both modifications, the relationship between pharmacokinetic parameters and covariates is no longer valid on day D_b.

   b) the changes in parameters are related to the modifications of the covariates. The conditional distribution of the parameters given the covariates must be the same on the two days. One can predict the parameters values on day D_b using the joint distribution estimated on day D_a and the new value of the covariates on day D_b.

SIMULATED EXAMPLE

In order to illustrate the proposed approach we have built a very simple simulation example. The only pharmacokinetic parameter of the model is the clearance CL of a drug given in continuous infusion. At steady state the concentration at any time is given by $C(t) = R_0/CL$ where R_0 is the rate of administration. Assume that $R_0 = 10$ and that the measurement error on $C(t)$ is Gaussian, with zero mean and a standard deviation σ_e proportional to the concentration: $\sigma_e = 0.05C$.

On day D_a, assume that CL is Gaussian with mean $m = 6$ and standard deviation $\sigma = 1.3$. Using a random number generator, the clearance value of 100 individuals were generated and the corresponding concentrations were then corrupted by the measurement error. A sample of 100 individuals with one concentration measurement was generated.

On day D_b, several simulation settings were used to generate the clearance values of the same 100 individuals:

1. The concentrations are assumed to be equal on days D_a and D_b. This ideal case was defined only to *test* the approach.

2. The clearances are assumed to be identical on the two days, but a new sample of concentrations (because of the measurement error) was generated.

3. & 4. Respectively small and important random changes in the clearance values are defined:

$$CL^b = CL^a + \Delta CL \tag{9}$$

where ΔCL is Gaussian with mean 0.5 (respectively 2) and standard deviation 0.25 (respectively 1). These random changes on clearance correspond to a mean increase of about 10% and 30%, respectively.

5. The individuals are assumed to be independent between the two days. A new sample of 100 clearances was drawn from the distribution of day D_a.

In the five cases, the likelihood of the two sets of concentration data on day D_a and D_b were maximized using the NPML method under the three hypotheses, H_0, H_1 and H_0' defined in the previous paragraph. These estimated values are reported in Table I.

These results illustrate that when the parameters are the same (case 2 in Table I) or when the data are identical (case 1), the likelihoods under H_0 and H_1 are equivalent, suggesting stationarity of the parameters. However, in case of new clearance values (case 5) or when there are important changes or even small changes (case 3), the likelihood under H_0 is substantially smaller than under H_1, indicating nonstationarity. The likelihoods under H_0' and H_1 are, as expected, very similar when the clearances at day D_b and D_a are independent (case 5). Furthermore, comparing L_0 and L_0' in cases 3 and 4 suggests that it is equivalent to assume that the parameters are stationary or independent when ΔCL is small. However, in case of important changes they show that it is better to assume that the parameters are independent, i.e., assume that measurements on the two days are from different individuals.

As a second example, the individual body weight was incorporated as a covariate in the simulation. It is assumed to be Gaussian with mean 60 and standard deviation 10. The clearance is related to body weight as follows: the clearance is assumed to be Gaussian with standard deviation 1 but with a mean proportional to body weight ($m = 0.1W$). For each individual, the body weight is first selected and then the clearance value is obtained from the resulting distribution.

On day D_b, modifications of body weight and clearance were simulated. First, changes were generated independently. The random increases in body weight and clearance, are Gaussian with mean and standard deviation 5 and 0.5, respectively. In a second simulation, the changes are related. The random increase in body weight ΔW is kept as before, but the increase in CL is $0.1\Delta W$. In both cases the mean increase in clearance is around 10%.

The estimated marginal distribution and conditional distribution of CL on body weight on day D_a and in each simulation settings on day D_b are displayed in Fig. 1. The corresponding estimated mean and standard deviations are reported on Table II.

These results illustrate that the estimated marginal distribution at day D_a has a greater dispersion than the conditional distribution on body weight. This is no longer the case on day D_b when changes are independent. Furthermore, it can be seen that the estimated conditional distributions under three different values of body weight appear similar at day D_a and day D_b when related changes are simulated;

Table I: Comparison of the Logarithm of the Maximum Likelihood Estimates from the Concentrations Measured at Days D_a and D_b Under the Three Hypotheses Using Various Simulation Settings for Clearance Evolution

Simulation Settings	H_1	H_0	H_0'
1: Same Data	79	78	-82
2: Same Parameters	47	43	-89
3: Small Changes	18	-93	-81
4: Important Changes	-18	-1380	-125
5: New Individuals	-100	-1037	-103

Table II. Estimated Mean and Standard Deviation (SD) of the Marginal Distribution of CL and the Conditional Distributions on Various Body Weights on Day D_a and on Day D_b

	Day D_a		Day D_b Independent		Day D_b Related	
Body Weight	Mean	SD	Mean	SD	Mean	SD
55	5.55	0.89	5.87	1.19	5.60	0.82
65	6.39	0.91	6.46	1.26	6.41	0.87
75	7.56	0.81	7.26	1.20	7.33	0.78
Marginal	5.90	1.30	6.49	1.43	6.40	1.25

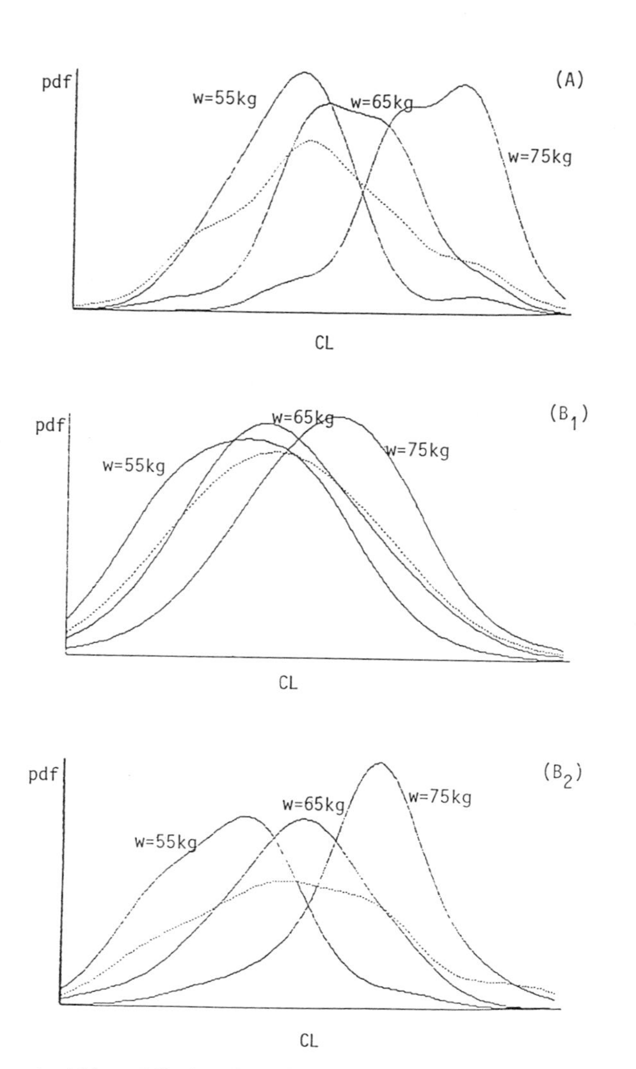

Fig. 1. Smoothed maginal (dotted line) and conditional distribution of CL on body weight estimated on day D_a (A), on day D_b with independent changes (B_1) and related changes (B_2).

their means and standard deviations are equivalent. When changes are independent, the conditional means at day D_b increase less with body weight than at day D_a and the standard deviations of the conditional distributions are larger.

APPLICATION TO ZIDOVUDINE POPULATION PHARMACOKINETICS

The problem of the nonstationarity of pharmacokinetic parameters was raised while studying AZT (formerly zidovudine) population characteristics. The high interindividual pharmacokinetic variability of this drug has already been reported [7, 8]. Thirty six AIDS patients were given 500mg of AZT every 6 hours during one month and then 250mg of AZT. Four blood samples were taken at 0 (predose), 0.5, 1 and 3 hours after the second dose on the first day and on the 35th day of therapy. Urine samples were collected on day 1 and day 35 during 24 hours. The plasma concentrations and amounts excreted in urine of AZT and its 5′-glucuronirated metabolic GAZT were assayed by HPLC [9].

A classical compartmental model (Fig. 2) was used to describe these data. It involves 5 parameters: the volumes of distribution of AZT and GAZT and the three rate constants k_m, k_e and k_{em}. However, a complex oral absorption model with two phases has been developed using well-documented kinetic data originating from another trial. Let D be the administered dose and F the bioavailability, the total dose absorbed is $F \cdot D$. At time 0, a fraction F' of $F \cdot D$ is supposed to be absorbed at a constant rate k_a; after a lag-time T_{lag}, the remaining fraction $(1 - F')$ is assumed to be absorbed instantaneously. The nine parameters of the model are, therefore, $(V_{AZT}, V_{GAZT}, F, F', k_m, k_e, k_{em}, T_{lag}, k_a)$.

The measurement error was assumed to be Gaussian, independent with zero mean. The variance on one measurement y is given by:

$$\sigma^2 = a + by^2 \tag{10}$$

with $a = 7$ for AZT and 30 for GAZT and $b = 0.2$ for plasma concentrations and 0.3 for urinary amounts.

A population approach was necessary to estimate the parameter values. The previous approach was used to study the nonstationarity of the parameters. The logarithm of the estimated likelihoods of the whole set of data was -1328 under the assumption on stationarity (H_0) and - 457 otherwise (H_1).

In a second analysis, four covariates measured at day 1 and 35 were incorporated: body weight, serum hemoglobin, bilirubin and creatinine. Unfortunately, no

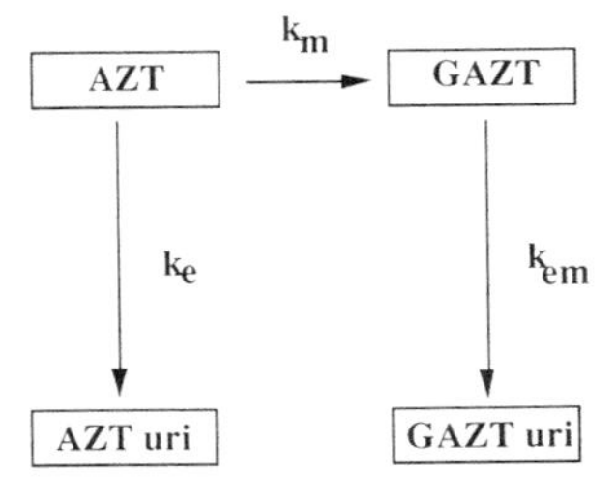

Fig. 2. The compartmental pharmacokinetic model of AZT (zidovudine) and GAZT in plasma and urine.

clear relationship between the pharmacokinetic parameters and the covariates was found. The covariates have poor predictive performance. For instance, the Bayesian estimates of the parameters based on the concentration data and covariates at day 1 and on the estimated distribution at day 1 lead to a root mean squared relative error (RMSRE) on AZT plasma concentration of 0.20 (this value is consistent with the variance of the error model). When only the covariates are used to estimate the parameters this RMSRE increases to 2.18.

Using the parameters estimated from both concentration and covariates at day 1 to predict the AZT concentration at day 35, i.e., assuming the stationarity of the parameters, gives a RMSRE of 2.44. However, assuming that changes in parameters are related to changes in covariates, i.e., estimating the parameters at day 35 from the new values of the covariates, produces a similar RMSRE of 2.23. It falls to 0.51 when both the covariates and the concentrations measured at day 35 are used. These results strongly suggest that zidovudine pharmacokinetic parameters are not stationary during therapy. The modifications of the pharmacokinetic parameters do not seem to be related to the modifications of the four covariates included in the analysis.

CONCLUSION

We have proposed a new method to check the stationarity of individual pharmacokinetic parameters from few individual data. It is based on a population analysis and no *a priori* assumption of the distribution of the parameters has to be defined. However, it does require a complete specification of both the pharmacokinetic model and the measurement error model.

We also proposed to test another null hypothesis of independence between the two days when the parameters appear to be nonstationary. Unfortunately no statistical test to reject or accept these null hypotheses has been defined. The simulated example, however, illustrates that the likelihoods associated with the different hypotheses appear to be very different even in the case of small modifications of the parameters. It must be noticed that when the parameters are not stationary, the joint distribution estimated under H_1 allows prediction, for a new individual, of the parameters on day D_b from information on day D_a.

We also proposed an approach to study the relationship between modifications of the parameters and of some biological covariates, in the case when the parameters are not stationary. Again, no assumption has to be defined on the relation between the parameters and the covariates. This approach remains mainly intuitive. The future is to build a population model where evolution of both parameters and covariates could be incorporated. However, when the number of parameters, covariates or times increase, this approach which is totally non-parametric may be limited and some parametric assumptions would perhaps have to be defined.

REFERENCES

1. B. Whiting, A. Kelman, and J. Grevel. Population pharmacokinetics. Theory and clinical application. *Clin. Pharmacokinet.* **11**:387-401 (1986).
2. J. L. Steimer, A. Mallet, and F. Mentré. Estimating interindividual pharmacokinetic variability. In M. Rowland, L. B. Sheiner, and J. L. Steimer (eds.), *Variability in Drug Therapy*, Raven Press, New York, 1985, pp. 65-111.
3. L. B. Sheiner and S. L. Beal. Bayesian individualization of pharmacokinetics: simple implementation and comparison with non-Bayesian methods. *J. Pharm. Sci.* **71**:1344-1348 (1982).

4. A. Mallet. A maximum likelihood estimation method for random coefficient regression models. *Biometrika* **73**:645-656 (1986).
5. A. Mallet, F. Mentré, J. L. Steimer, and F. Lokiec. Nonparametric maximum likelihood estimation for population pharmacokinetics. An application to Cyclosporine. *J. Pharmacokin. Biopharm.* **16**:311-327 (1988).
6. A. Mallet, F. Mentré, J. Gilles, A. W. Kelman, A. N. Thomson, S. M. Bryson, and G. Whiting. Handling covariates in population pharmacokinetics with an application to gentamicin. *Biomed. Meas. Infor. Contr.* **2**:673-683 (1988).
7. J. M. Collins and J. D. Unadkat. Clinical pharmacokinetics of zidovudine. An overview of current data. *Clin. Pharmacokinet.* **17**:1-9 (1989).
8. S. R. Gitterman, G. L. Drusano, M. J. Egorin, H. C. Standiford, and the Veterans Administration Cooperative Studies Group. Population pharmacokinetics of zidovudine. *Clin. Pharmacol. Ther.* **48**:161-167 (1990).
9. S. S. Good, D. J. Reynolds, and P. de Miranda. Simultaneous quantification of zidovudine and its glucuronide in serum by high-performance liquid chromatography. *J. Chromatogr.* **431**:123-133 (1988).

ON THE SINGLE-POINT, SINGLE-DOSE PROBLEM

George D. Swanson

Anesthesiology Department
University of Colorado Medical School
and
Department of Physical Education
California State University, Chico

INTRODUCTION

The problem of predicting dosage requirements for individual patients has received considerable attention in recent years. The use of average values of pharmacokinetic parameters for predicting dosage rates or steady-state drug concentrations has obvious limitations due to inter-patient variability. In order to incorporate individual variability into such predictions, the "single-point, single-dose" method was introduced by Slattery *et al.* [1]. In this method, one measurement of serum drug concentration is made at a sampling time, t_s, after a test dose is given. This single measurement is then used in a linear, one compartment model to predict relevant pharmacokinetic variables. Slattery *et al.* [1], as well as other investigators [2–4], have indicated an optimal sampling time of $t_s = k_o^{-1}$ where k_o is the average population value of the elimination rate constant. These analyses generally utilize a localized linearization argument, which is strictly valid only in the limit as the population variance of the rate constant approaches zero.

This present paper extends the work of these previous studies by assigning specific statistical distributions to the elimination rate constant and characterizing the resulting distribution of the variable of interest via analytical parameters or computer simulation results. Optimal sampling times are then determined from global characteristics. These results suggest that $t_s < k_o^{-1}$ when the variance of the rate constant is appreciably different from zero. Furthermore, sampling at $t_s = k_o^{-1}$ for these large variance problems can result in appreciable overdosing for individual subjects.

METHODS

The methods of this paper will follow those of our previous work [5]. Consider the situation in which a target drug concentration C_T is specified and is to be maintained by a constant infusion rate. Using a linear, one-compartment model, the serum drug concentration at time t is given by

$$C(t) = \frac{D}{Vk}(1 - e^{-kt}), \qquad t > 0 \tag{1}$$

Advanced Methods of Pharmacokinetic and Pharmacodynamic Systems Analysis
Edited by D'Argenio, Plenum Press, New York, 1991

where D is the dose rate (mass/time), V is the volume of distribution, and k is the elimination rate constant. The steady-state concentration given by this model is $C_T = D/Vk$. An initial estimate for the required infusion rate to maintain the target level is simply $D = C_T V k$ where, in the absence of any additional information, average population values of V and k would be used.

In order to obtain additional, patient-specific information, we allow one measurement of serum drug concentration at time t_s following an injection of a test dose d. This measured concentration (C_s) is given by

$$C_s = \frac{d}{V} e^{-kt_s} \tag{2}$$

This measurement is assumed to be made without error and may be used to determine the volume of distribution V in terms of the other parameters. An improved estimate of the required infusion rate then follows by solving Eq. (2) for V and substituting into the equation for D:

$$D = C_T V k = \frac{C_T d k e^{-kt_s}}{C_s} \tag{3}$$

At this point, t_s is arbitrary and the only value of k available is an average, population value.

It is recognized that the elimination constant k has some variability across all patients. This variability in k produces uncertainty in the infusion rate D. While sampling time t_s is arbitrary, there is a choice of t_s that minimizes the variation in D due to the variation in k.

It is now assumed that the rate constant k has a known distribution. We shall investigate an optimal dosing strategy along with the resulting distribution of the steady-state drug concentrations. This problem is formulated in the following way.

Each patient has an individual elimination rate constant and volume of distribution that are denoted by k_i and V_i, respectively. If these two quantities were known, then an individual infusion rate $D_i = C_T k_i V_i$ could be prescribed to reach the target concentration C_T. One measurement is allowed following a test dose in order to obtain additional information. Specifically, this measurement of concentration C_s taken at sampling time t_s can be used to estimate the individual volume V_i so that

$$V_i = \frac{d}{C_s} e^{-k_i t_s} \tag{4}$$

This leads to an infusion rate given by

$$D_i = \frac{C_T d}{C_s} k_i e^{-k_i t_s} \tag{5}$$

In practice, the value of the elimination rate constant for the ith patient is not known. Therefore, the average value (k_o) would be used yielding an estimate of the infusion rate as

$$\widehat{D_i} = \frac{C_T d}{C_s} k_o e^{-k_o t_s} \tag{6}$$

The resulting serum blood concentration for the ith patient is

$$C_i = \frac{\widehat{D}_i}{V_i k_i} = \frac{C_T d}{C_s V_i k_i} k_o e^{-k_o t_s} \tag{7}$$

where C_s is the measured concentration for the ith patient given by

$$C_s = \frac{d}{V_i} e^{-k_o t_s} \tag{8}$$

Combining Eqs. (7) and (8) yields the concentration ratio

$$r = \frac{C_T}{C_i} = \frac{k_i}{k_o} e^{-t_s(k_i - k_0)} = f(k_i) \tag{9}$$

This ratio has the useful property that if a patient is *average* with $k_i = k_o$, then $r = 1$. Furthermore, a deviation away from $r = 1$ implies an error in the delivered concentration. Minimizing the variance in r minimizes the error in delivered concentrations. Note that the maximum r can easily be determined as

$$r_{max} = 1/\left(k_o t_s e^{(1-k_o t_s)}\right) \tag{10}$$

In contrast, the minimum r depends upon the distribution for the rate constant k.

At this point, we could impose any sort of distribution upon k. As a first approach, it will be assumed that k is Normally distributed about a known mean k_o with a known variance σ^2. Therefore, the probability density function for k may be written as

$$\phi(k) = \frac{1}{\sigma\sqrt{2}\,\pi} e^{-[(k-k_0)^2/2\sigma^2]} \tag{11}$$

The problem of finding the resulting density function of $r = f(k)$ is solvable in some cases. However, a solution cannot be given in closed form unless the inverse of f can be written explicitly. If this cannot be done, as in the present case, then the density function of r must be determined numerically [5].

Even without an explicit representation for the density function of r, it is possible to make some progress toward determining the sampling time t_s which minimizes the variance of r. Proceeding analytically [5], the variance of r is given by

$$\text{var}(r) = \frac{1}{k_o^2} e^{\sigma^2 t_s^2} (e^{\sigma^2 t_s^2} (\sigma^2 + k_o^2 - 4 t_s k_o \sigma^2 + 4\, t_s^2 \sigma^4) - k_o^2 + 2 t_s k_o \sigma^2 - t_s^2 \sigma^4) \tag{12}$$

The concentration ratio r was formulated above as C_T/C_i so as to avoid the mathematical problems associated with $k_i = 0$. A more appropriate formulation would be to define $r_1 = C_i/C_T$ so that $r_1 > 1$ reflects an overdose and $r_1 < 1$ reflects an underdose. Our second approach then formulates the problem with the inverse concentration ratio r_1 as:

$$r_1 = \frac{C_i}{C_T} = \frac{k_o}{k_i} e^{t_s(k_i - k_o)} = f_1(k_i) \tag{13}$$

To insure that $k_i > 0$ and reflect a more realistic situation, the distribution for k_i is given by a lognormal distribution. That is, the previous Normal distribution for k is transformed by an exponential function (see below)

The analytical properties for r_1 are not tractable. Therefore, computer simulation results were obtained using MINITAB to generate numerical lognormal distributions for k_i. These numerical distributions were then utilized with Eq. (13) to yield numerical distributions for r_1. The range and standard deviation for r_1 were determined numerically as a function of sampling time t_s.

In our previous work $k_o = 1$ and $\sigma = 0.2$, which yields $t_s \approx 1$ as the optimal sampling time [5]. In this present paper, $k_o = 1$ and $\sigma = 0.6$ for the Normal distribution of rate constants. This distribution was then transformed by $e^{0.89(\cdot)}/2.7$ to yield a distribution of rate constants, which has a lognormal distribution with a mean of approximately one and a standard deviation of approximately 0.6.

RESULTS

The following results pertain to these large variation rate constant distributions with mean $k_o = 1$ and standard deviation $\sigma = 0.6$. To explore the optimal sample time problem with the concentration ratio given by r in Eq. (9), the rate constant distribution was assumed to be Normal and thus, the analytical result for variance of r given in Eq. (12) was used. To explore the optimal sample time problem with the inverse concentration ratio given by r_1 in Eq. (13), the rate constant distributions were assumed to be lognormal with the simulation package MINITAB used to obtain results.

Figure 1 indicates the variance of the concentration ratio r_1 as a function of sampling time t_s. Note that when $\sigma = 0.2$ the optimal sampling time occurs at one as predicted from the small variance, localized argument. However, when $\sigma = 0.6$, the optimal sampling time occurs as t_s approaches 0.5. This implies for the large variance problem that $t_s < k_o^{-1}$.

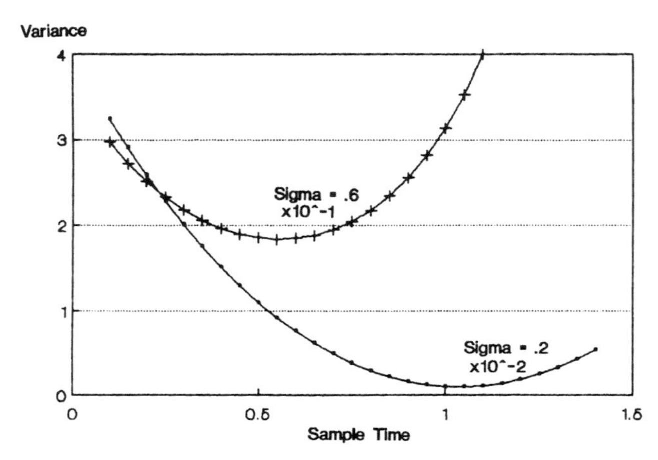

Fig. 1. Variance as given by Eq. (12) as a function of sampling time. Note the optimal sample time when $\sigma = 0.2$ is 1. In contrast, when $\sigma = 0.6$, the optimal sample time is 0.5–0.6.

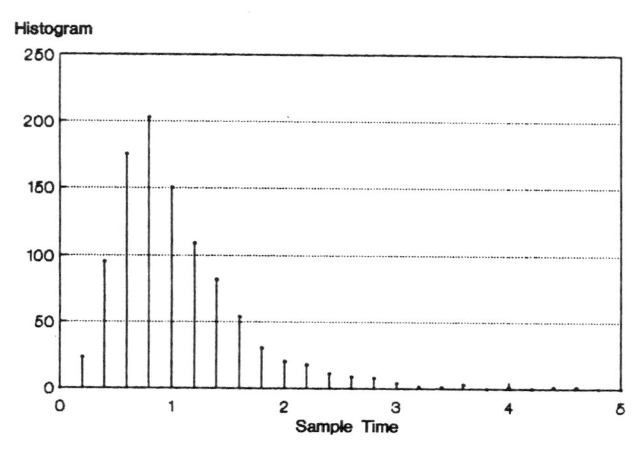

Fig. 2. Histogram of typical lognormal distribution generated via MINITAB.

This result is also predicted by the distribution for the inverse concentration ratio r_1 when the rate constant distribution is lognormal. Figure 2 indicates a typical lognormal distribution generated via MINITAB. Note that all rate constants are greater than zero with the distribution characterized by a positive skew.

Figure 3 plots the inverse concentration ratio r_1 (Eq. (13)) as a function of rate constant. Note how the shape of the plot becomes less curved as the sampling time is decreased from $t_s = 1.5$ to $t_s = 1.0$ to $t_s = 0.5$. These plots represent "transfer curves" for the lognormal distribution of rate constants shown in Fig. 2. The flatter

Table I. Characteristics of the Inverse Concentration Ratio r_1 Distribution for Six Random Realizations of Lognormal Rate Constant Distributions. Note that the Minimum Values are Consistent with Eq. (10); that is, $r_{1min} = k_o t_s e^{(1-k_o t_s)}$.

Trial	Sample Time	Mean	Std. Dev.	Min	Max
1	1.00	1.210	0.522	1.000	11.200
	0.75	1.840	0.316	0.963	4.090
	0.50	1.200	0.424	0.824	4.360
2	1.00	1.205	0.650	1.000	16.002
	0.75	1.171	0.321	0.963	5.231
	0.50	1.179	0.401	0.824	4.562
3	1.00	1.038	0.584	1.000	4.272
	0.75	1.167	0.277	0.963	3.436
	0.50	1.182	0.398	0.824	4.245
4	1.00	1.183	0.374	1.000	6.527
	0.75	1.157	0.261	0.963	3.376
	0.50	1.164	0.371	0.824	4.167
5	1.00	1.198	0.508	1.000	11.501
	0.75	1.183	0.318	0.963	4.163
	0.50	1.204	0.426	0.824	3.871
6	1.00	1.193	0.378	1.000	5.422
	0.75	1.168	0.288	0.963	3.214
	0.50	1.176	0.409	0.824	3.958

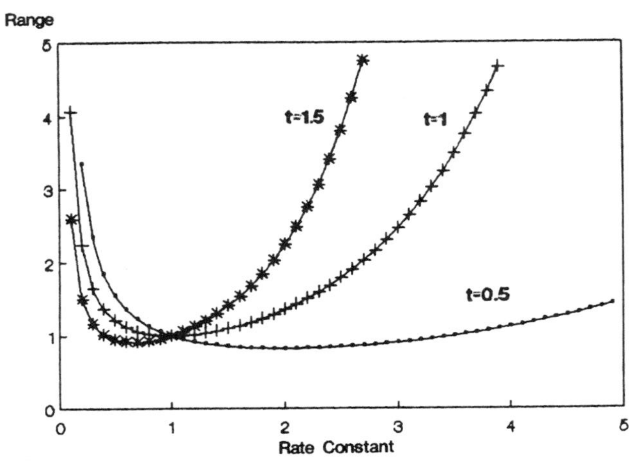

Fig. 3. Transfer curves for r_1 as a function of rate constant. Note that when $t_s = 0.5$, the transfer curve has the least curvature indicating a concentration of probability mass around one.

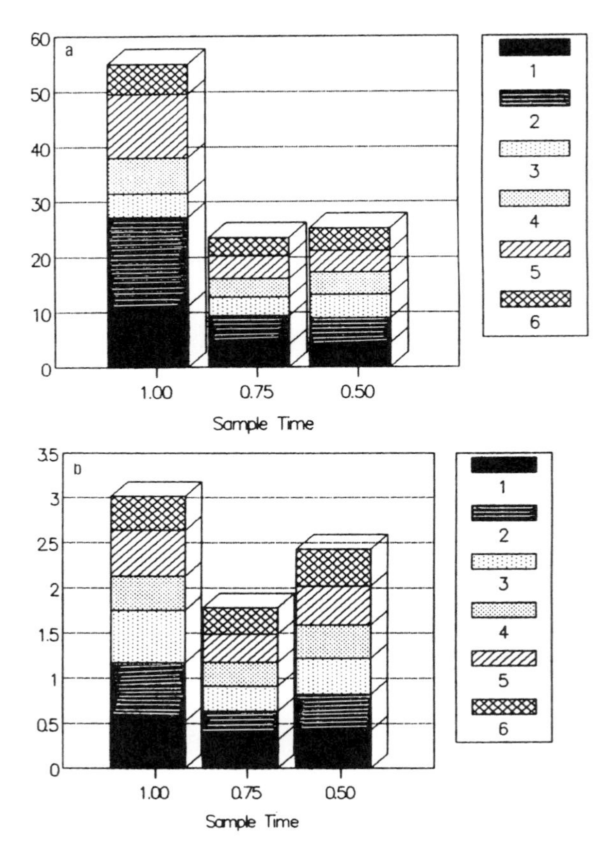

Fig. 4. Distribution characteristics for r_1 as a function of sample time. The results of six random realizations for the lognormal distribution of Fig. 2 are shown. Upper panel (a) shows the maximum values, while the lower panel (b) shows the standard deviations.

the transfer curve the more the probability mass for r_1 is "concentrated" around $r_1 = 1$. Thus, sampling times of $t_s = 1.5$ and $t_s = 1.0$ transfer marked amounts of probability mass to large r_1 values resulting in major overdose for these patients. In contrast, the transfer curve for $t_s = 0.5$ transfers this probability mass to r_1 values around one. This implies that the optimal sampling time $t_s < k_o^{-1}$.

Table I tabulates the characteristics of the distribution of the inverse concentration ratio r_1 for six different random realizations of the lognormal distribution shown in Fig. 2. Note that in general, the r_1 distribution characteristics exhibit the highest bias, standard deviation and range when the sampling time is $t_s = 1.0$. When the sampling time is reduced, these characteristics tend to be reduced. Note that $t_s = 0.75$ yields a substantially lower standard deviation and range when compared to the $t_s = 1.0$ values. Also, $t_s = 0.75$ yields marginally better distribution characteristics than $t_s = 0.5$. These tabulated results are shown schematically in Fig. 4.

DISCUSSION

These results suggest that for large variation rate constant distributions ($\sigma \approx 0.6k_o$) the optimal sample time is given by $t_s < k_o^{-1}$. Furthermore, if $t_s = k_o^{-1}$, patients with large rate constants would receive maintenance doses that would produce substantial overdoses. Therefore, these results have major clinical implications. These concepts are also evident in the recent results of Bahn and Landaw [6].

The problem of optimal sample time as formulated here assumed a measurement of blood plasma drug concentration that was error free. The measurement error problem still remains to be explored. However, heuristically, as measurement noise is added, the sampling time would decrease so as to maintain measurement signal to noise ratio high. This also suggests that $t_s < k_o^{-1}$ when measurement noise is present.

REFERENCES

1. J. T. Slattery, M. Gibaldi, and J. R. Koup. Prediction of drug concentration at steady state from a single determination of concentration after an initial dose. *Clin. Pharmacokin.* **5**:377–385 (1980).
2. M. Dossing, A. Volund, and H. E. Poulsen. Optimal sampling times minimum variance of clearance determination. *Brit. J. Clin. Pharmaco.* **15**:231–235 (1983).
3. J. R. Koup. Single-point prediction methods: A critical review. *Drug Intel. Clin. Pharm.* **16**:855–862 (1982).
4. J. D. Unadkat and M. Rowland. Further considerations of the single-point single-dose method to estimate individual maintenance dosage requirements. *Ther. Drug Monit.* **4**:201–208 (1982).
5. W. L. Briggs, R. W. Phelps, and G. D. Swanson. A probabilistic approach to the single-point, single-dose problem. *IEEE T. Bio-med. Eng.* **37**:80–84 (1990).
6. M. M. Bahn and E. M. Landaw. A minimax approach to the single-point method of drug dosing. *J. Pharmacokin. Biopharm.* **15**:255–269 (1987).

APPLICATION OF STOCHASTIC CONTROL THEORY TO OPTIMAL DESIGN OF DOSAGE REGIMENS

Alan Schumitzky

Department of Mathematics
University of Southern California

ABSTRACT

Designing a dosage regimen for a pharmacokinetic/pharmacodynamic system involves defining: i) a patient-dependent model, which includes structure, parameter, and measurement uncertainties; ii) the choice of controls, which can include dose amounts, dose times and/or sampling times; and iii) an appropriate performance index to evaluate achievement of a clinically chosen therapeutic goal. The control problem then is to choose the dosage regimen that optimizes the expected value of the performance index. This problem fits within the framework of stochastic control theory. Examples are given to illustrate the variety of this class of problems, including: optimal dose regimens for target level and target window cost; and optimal sampling schedules for maximal information. By varying the class of admissible controls, different strategies are generated. Control strategies to be discussed include: open loop, open loop feedback, separation principle, and iteration in policy space. Monte Carlo simulation studies of a terminal cost type problem are presented.

INTRODUCTION

Designing a dosage regimen for a pharmacokinetic/pharmacodynamic (PK/PD) system involves defining:

- a patient-dependent PK/PD model, which includes structure, parameter, and measurement uncertainties;
- the choice of "controls", which can include dose amounts, dose times and/or sampling times; and
- an appropriate performance index to evaluate achievement of a clinically chosen therapeutic goal.

The control problem then is to choose the dosage regimen that optimizes the expected value of the performance index. This problem fits within the framework of "stochastic" control theory, i.e., control in the presence of uncertainty.

Examples are given to illustrate the variety of this class of problems in the PK/PD setting. Included will be: optimal dose regimens for target level and target window cost; and optimal sampling schedules for maximal information.

Advanced Methods of Pharmacokinetic and Pharmacodynamic Systems Analysis
Edited by D'Argenio, Plenum Press, New York, 1991

By varying the class of admissible controls, different strategies are generated. Control strategies to be discussed include: open loop, open loop feedback, separation principle and iteration in policy space. Monte Carlo simulation studies of a terminal cost type problem are presented.

STOCHASTIC CONTROL FORMALISM

In this section we define the various ingredients which make up a stochastic control problem.

State Equations

The time history of a drug concentration in a traditional PK model is typically described by a system of deterministic ordinary differential equations. When "noise" is added to this model the system of differential equations becomes "stochastic". Since stochastic differential equations are mathematically quite sophisticated, it is desirable to seek a simpler formulation. The possibility of such a simplification comes from the fact that dose inputs and infusion rates are changed only at discrete time points (as opposed to continuously). In the linear case, this leads to an equivalent linear system of "discrete time" stochastic equations (as illustrated below). In the nonlinear case the appropriate model is given by a discrete time "Markov process" [1]. For purposes of this exposition, we take the middle ground and assume that the PK/PD model can be described by a nonlinear system of discrete time stochastic equations as follows:

$$x_{n+1} = f_n(x_n, u_n, w_n, \phi), \qquad n = 0, 1, \ldots, N \tag{1}$$

where at "stage" n, x_n is the "state" vector of the system, u_n is the external "control" vector, w_n is the "process noise" vector, and f_n is a known vector function. Further, ϕ is an unknown time invariant parameter vector.

In the PK/PD setting, the state x_n typically corresponds to amounts and concentrations of drug in various compartments and/or to various drug effects. The control u_n typically corresponds to the drug amounts and/or infusion rates of one or more drugs into various compartments; additionally, the control u_n could also include various design entities such as dose times and sample times. The parameter vector ϕ typically corresponds to various PK entities such as rate constants, distribution volumes, clearances, and/or to various PD entities such as C_{50} and E_{max}. The process noise vector w_n typically corresponds to various errors made in dose amount, dose timing and model mispecification. Finally, the stage n typically corresponds to a dose time or infusion time or sample time.

Measurement Equations

The measurement model is also described by discrete time stochastic equations. These equations can be written as :

$$y_n = h_n(x_n, u_n, v_n, \phi), \qquad n = 1, 2, \ldots, N \tag{2}$$

where at stage n, y_n is the "measurement" vector ; v_n is the corresponding "measurement noise" vector, and h_n is a known vector function.

In the PK/PD setting, the measurement y_n typically corresponds to concentrations of drug in serum or amounts of drug in urine but could also be measured drug effects; the measurement noise v_n typically corresponds to assay or measurement device noise but could also correspond to the errors in recording the time of observations.

Prior Distributions

In the model of Eqs. (1) and (2), the vectors x_0, w_n, v_n, and ϕ are the basic random variables; and the prior probability density functions:

$$p(x_0 \mid \phi), \qquad p(w_n \mid x_n, u_n, \phi), \qquad p(v_n \mid x_n, \phi), \qquad p(\phi) \tag{3}$$

are assumed known. All the other random vectors x_n, u_n, and z_n are functions of x_0, w_n, v_n, and ϕ.

The determination of these prior distributions is an important ingredient in the stochastic control formulation. For the measurement noise, the prior distribution of v_n (given x_n and ϕ) comes from "calibrating" the measurement devices. For the PK/PD parameters, the prior distribution of ϕ comes from analyzing previous studies. This latter problem is called "Population Analysis" and is the subject of much interest in PK/PD applications. (See [2, 3] for survey articles and the chapter by Mentré and Mallet in this volume.) The initial state vector x_0 is quite often known exactly (e.g., $x_0 = 0$). Otherwise the prior distribution of x_0 must be determined from "prior" events. On the other hand, the process noise term w_n is a relatively new addition to PK/PD problems. The determination of its prior distribution (given x_n, u_n and ϕ) is, for all practical purposes, an essentially unexplored problem. (See [4, 5] and the chapter by D'Argenio in this volume.)

Admissible Controls

A realizable control u_n must be *nonanticipatory*. This means that u_n can only depend on present and past data. More precisely:

$$u_n \text{ is a function of the information } I_n = (y_1, \ldots, y_n; u_0, \ldots, u_{n-1}). \tag{4}$$

For our PK/PD applications, the components of u_n will be constrained (doses cannot be negative or arbitrarily large, dose times must be sequential, etc.). Further, it may also be necessary to constrain certain components of the state x_n (serum levels should not be too large, platelet counts should not be too small). Since x_n is random, these constraints will only be required to hold in a probabilistic sense. Such constraints also (implicitly) imply constraints on the controls $(u_0, u_1, \ldots, u_n)$. All these constraints will be collected under the assumption:

$$u_n \text{ belongs to some set } \mathcal{U}_n \text{ which may depend on } (u_0, u_1, \ldots, u_{n-1}). \tag{5}$$

A control *policy* $U \equiv (u_0, u_1, \ldots, u_N)$ is called *admissible*, if U and the resulting $X \equiv (x_0, x_1, \ldots, x_{N+1})$ satisfy Eqs. (1)–(5). The control problem then is to choose the admissible control policy U to maximize some "performance index" or minimize some "cost function".

For example it may be required to design a regimen to maximize the probability that certain drug levels and/or effects belong to some therapeutic window. Here, an appropriate performance index would be of the form:

$$J(U) = \sum_{n=0}^{N} \alpha_n \, \text{Prob}\{g_n(x_n) \in S_n\} \tag{6}$$

where $g_n(x)$ is a given function of x, and S_n is a given set (the therapeutic window). In this case U would be chosen to maximize $J(U)$.

Similarly, it may be required to design a regimen to minimize the error between certain drug levels effects and some desired response. Here an appropriate cost function would be of the form:

$$J(U) = \mathrm{E}\{\sum_{n=0}^{N} \alpha_n[\| x_{n+1} - L_{n+1} \|^2 + \gamma_n \| u_n \|^2]\} \tag{7}$$

where L_n is the desired response at stage n, $\| \cdot \|$ is some vector "norm", and where α_n and γ_n are given nonnegative constants reflecting the relative importance of the corresponding terms (including, for example $\gamma_n \equiv 0$). In this case U would be chosen to minimize $J(U)$.

Finally, to illustrate an example that is not normally considered in the context of stochastic control, it may be required to design a sampling schedule to maximize some index of "information". Here, an appropriate performance index could be of the form:

$$J(U) = E\,\{\det\,[M(\phi, U)]\} \tag{8}$$

where

$$M(\phi, U) = E\left\{(\partial \log p(Y \mid \phi)/\partial\phi)^T(\partial \log p(Y \mid \phi)/\partial\phi) \mid \phi\right\} \tag{9}$$

is the Fisher information matrix and where $Y = (y_1, y_2, \ldots, y_N)$. In this case U would be chosen to maximize $J(U)$.

These three examples can be put into the following form: Some criterion function $C = C(U, X, \phi)$ is given and the expected value of C is to be optimized. To be specific, we will suppose that $E\{C\}$ is to be minimized. (For maximization problems, just replace C by $-C$.)

Further, for technical reasons, it will be assumed that $C(X, U, \phi)$ is of the form:

$$C(U, X, \phi) = \sum_{n=0}^{N} g_n(x_{n+1}, u_n, \phi) \tag{10}$$

where g_n is the "cost" at stage n. (It is clear that the criterion functions in Eqs. (6),(7) are in the form of Eq. (10). To put the criterion function in Eq. (9) into this form requires "additional" state variables. This is illustrated in Example 2.)

The stochastic control problem can now be stated precisely as follows. Find the admissible policy $U^* = (u_0^*, u_1^*, \ldots, u_N^*)$ which minimizes

$$J(U) = E\{C(U, X)\} \tag{11}$$

over all admissible U.

CONTROL POLICIES

The stochastic control problem as stated above is extremely general. It can be made to reflect many of the decisions made in a clinical situation. Unfortunately, the optimal control U^* cannot be implemented except for the most trivial cases. The reason for this is outside the scope of this paper, but it essentially goes under the name of "The Curse of Dimensionality". In this section we, therefore, discuss various types of suboptimal control policies which can be implemented. We restrict our attention to those policies which have appeared in PK/PD settings.

Open Loop

In the *open loop* policy, the controller ignores all measurement data and depends only on the prior "information" $I_0 = \{p(x_0 \mid \phi),\ \ p(w_n \mid x_0, \phi),\ \ p(\phi)\}$. The optimal open loop (OL) control $U^{OL} \equiv (u_0^{OL}, \ldots, u_N^{OL})$ minimizes the expression

$$E\left\{\sum_{i=0}^{N} g_i(x_{i+1}, u_i, \phi) \mid I_0\right\} \tag{12}$$

with respect to the deterministic sequence $(u_0, u_1, \ldots, u_N)$. (For notational simplicity, in Eq. (12) and throughout this section, we assume that $g_n(x_{n+1}, u_n, \phi) \equiv \infty$, if $u_n \notin \mathcal{U}_n$ so that the constraints of Eq. (5) can be suppressed.)

Open Loop Feedback

In the *open loop feedback* policy, the controller at stage n acknowledges that the information I_n is available, but assumes that no measurements will be taken in the future. The optimal open loop feedback (OLF) control $U^{OLF} \equiv (u_0^{OLF}, \ldots, u_N^{OLF})$ is such that, at each stage n, u_n^{OLF} minimizes the expression

$$\min_{u_{n+1}\cdots u_N} E\{g_n(x_{n+1}, u_n, \phi) + \sum_{i=n+1}^{N} g_i(x_{i+1}, u_i, \phi) \mid I_n\} \tag{13}$$

with respect to u_n.

In Eqs. (12) and (13), the expression $E\{C \mid I_n\}$ denotes the expectation of C conditioned on the information I_n. Note that the optimal OLF control is just the optimal OL control starting at stage n with "prior" information I_n. The fact that measurements can be used to advantage is reflected by the result [6]:

$$J(U^{OLF}) \leq J(U^{OL})$$

Separation Principle

In the *separation principle* policy, the controller applies at each stage the deterministic control that would be applied if all random terms were fixed at their expected values. The optimal separation principle (SP) control $U^{SP} \equiv (u_0^{SP}, \ldots, u_N^{SP})$ is such that, at each stage n, u_n^{SP} minimizes the deterministic expression:

$$g_n(x_{n+1}, u_n, \widehat{\phi}) + \min_{u_{n+1}\cdots u_N}\left[\sum_{i=n+1}^{N} g_i(x_{i+1}, u_i, \widehat{\phi})\right]$$

with respect to u_n, where

$$x_{i+1} = f_i(x_i, u_i, 0, \widehat{\phi}), \qquad x_i = \widehat{x}_n, \quad i = n, \ldots, N \tag{14}$$

In Eq. (14) it is assumed that $E\{w_i\} = 0$. Further, $\widehat{x}_n$ and $\widehat{\phi}$ are some estimates of x_n and ϕ based on the information I_n.

The separation principle is so named as the resulting controller "separates" the problem of estimation and control. The popularity of the SP controller comes from its ease of computation and from the fact that the optimal SP controller is actually optimal with respect to all admissible controls in the linear, quadratic, Gaussian case (LQG). In the LQG case the state and measurement equations are linear, the cost function is quadratic, all noise terms are independent Gaussian, there is no unknown parameter vector ϕ, and $\widehat{x}_n = E\{x_n \mid I_n\}$. Unfortunately these conditions rarely hold in PK/PD applications, and in general:

$$J(U^{SP}) \not\leq J(U^{OL})$$

Iteration in Policy Space

The optimal control policy $U^* = (u_0^*, u_1^*, \ldots, u_N^*)$ defined in Eq. (11) satisfies an important recursion relationship. At each stage n, the control u_n^* minimizes the expression

$$E\{g_n(x_{n+1}, u_n, \phi) + \sum_{i=n+1}^{N} g_i(x_{i+1}, u_i^*(I_i), \phi) \mid I_n\} \tag{15}$$

with respect to u_n. In Eq. (15), the dependence of u_i^* on I_i is explicitly shown to indicate the way that u_i^* depends on u_n, for $i = n+1, \ldots, N$. Equation (15) leads to Bellman's method of Stochastic Dynamic Programming. (This was one of the earliest and most important results in stochastic control theory [6, 7].)

The major drawback of using Eq. (15) for computational purposes is that it must be solved "backwards in time", since otherwise u_n^* would depend on the future controls $(u_{n+1}^*, \ldots, u_N^*)$. This problem led Bayard [8, 9] to consider analogous methods which could be solved "forwards in time". A brief description of Bayard's approach is as follows: Let $U^0 \equiv (\mu_0, \ldots, \mu_N)$ be any admissible control policy (called the *nominal policy*) . Then an *iteration in policy space* (IPS), with respect to U^0, is the control policy $U^{IPS} \equiv (u_0^{IPS}, \ldots, u_N^{IPS})$ such that, at each stage n, u_n^{IPS} minimizes the expression

$$E\{g_n(x_{n+1}, u_n, \phi) + \sum_{i=n+1} g_i(x_{i+1}, \mu_i(I_i), \phi) \mid I_n\} \tag{16}$$

with respect to u_n. It can be shown that U^{IPS} improves on U^0, i.e.,:

$$J(U^{IPS}) \leq J(U^0)$$

(Note the similarity between Eq. (15) and Eq. (16). Also note that calculating U^{IPS} from Eq. (16) is, in fact, simpler than calculating U^{OLF} from Eq. (13).)

This iteration process can be continued. The control policy U^{IPS} is admissible and can, therefore, be substituted for the original nominal policy. The resulting iteration on the policy U^{IPS}, now denoted by U^{2-IPS}, will, therefore, satisfy

$$J(U^{2-IPS}) \leq J(U^{IPS}).$$

A remarkable result of Bayard [8] is that if this process is continued for (at most) N iterations, then the optimal control U^* of Eq. (11) is obtained, i.e.,

$$U^{N-IPS} = U^*$$

However, even one iteration in policy space can dramatically improve control performance. This is illustrated in the simulation study presented below (Example 1).

Active versus Passive Learning

An important feature of the optimal control policy U^* is how it "learns" about the unknown parameters and states. In contrast to the OL, OLF, and SP controllers which learn only "passively", the optimal policy learns "actively" by probing the system. Probing comes from the anticipation that future measurements will be made, so that "mistakes" can be corrected [10]. In this sense, the optimal policy does experimental design "on line".

It is somewhat surprising that even one iteration in policy space on a "passive" nominal can generate a "active" controller. For example, it is shown in [11] that if the nominal policy is taken as U^{OLF} then the resulting U^{IPS} has this behavior. It is observed in the simulation study presented below (Example 1), that the same is true if the nominal policy is U^{SP}.

Previous PK/PD Applications

There have been only a few PK/PD applications of stochastic control theory to dynamical systems with process noise. In [12], an SP controller was simulated for a PK/PD problem in anticoagulant therapy; in [13] an OL controller was simulated for a PK problem in theophylline therapy; and in [14], an *IPS* type controller was simulated for a PK "terminal cost" problem.

Deterministic State Equations

In traditional PK/PD problems, the state Eq. (1) is deterministic (for given ϕ). That is, the process noise is assumed to be zero ($w_n \equiv 0$) and the initial state x_0 is known exactly, e.g., $x_0 = 0$. Now the cost function, Eq. (11), is of the form:

$$J(U) = \int C(U, X, \phi)p(\phi)d\phi$$

In this case the special policies OL, OLF, SP, and IPS above are all considerably simpler. Most previous PK/PD applications of stochastic control have been in this setting.

One of the earliest applications of "sophisticated" stochastic control theory was due to Gaillot, Steimer, Mallet, Thebault, and Bieder in 1979 where an optimal OL controller was utilized for lithium therapy [15] . In Richter and Reinhardt [16], a similar OL controller was utilized for theophylline therapy. More recently, Mallet, *et al.* [17] combined sophisticated population analysis with an optimal OL controller in designing dosage regimens for gentamicin therapy.

Simulated PK applications of optimal OLF control appeared in Katz and D'Argenio [18] and D'Argenio and Katz [19]. Applications of SP type controllers are more numerous. One of the earliest was the MAP Bayesian controller of Sheiner [20]. Similar controllers are found in the USC PC Pack of Jelliffe, *et al.* [21, 22]. Surveys and tutorials in this subject are found in Vozeh and Steimer [23], and Schumitzky [24, 25].

EXAMPLE 1: OPTIMAL INFUSION REGIMEN

In the next two sections we illustrate by example some of the ingredients of the stochastic control formalism. For this purpose it is sufficient to consider the simplest PK settings.

In this section we consider a one compartment model with iv infusion. In general, one of the problems with the "textbook" version of the stochastic control formalism is that it does not easily conform to the standard setting in PK/PD applications. The solution to this problem is a version of the stochastic control formalism which includes continuous time state equations and discrete time measurement equations. However, this is not the place for such a digression. It will be apparent in the example below that certain awkward constructions could be avoided by a "continuous-discrete" formalism.

State Equation

In the deterministic case, the time history of the drug concentration satisfies the differential equation:

$$\frac{dC(t)}{dt} = -kC(t) + \frac{r(t)}{V}, \quad t \geq 0, \qquad C(0) = 0 \tag{17}$$

where at time t, $C(t)$ is the concentration of drug, $C(0) = 0$; and $r(t)$ is the iv infusion rate which is assumed to be piecewise constant:

$$r(t) = r_n, \ t \in [t_n, t_{n+1}), \ \ n = 0, 1, \ldots, N,$$

where $\{t_n\}$ are the times at which the infusion rates can change. Further, k is the elimination rate constant and V is the volume of distribution.

If a Gaussian "white" noise process $W(t)$ with mean 0 and "variance" Q is added to Eq. (17), then the differential equation becomes stochastic and is written mathematically as:

$$dC(t) = -\left\{kC(t) + \frac{r(t)}{V}\right\} dt + d\beta(t), \qquad t \geq 0 \tag{18}$$

where $\beta(t)$ is a so-called Brownian motion such that $W(t) = d\beta(t)/dt$. (The noise term $W(t)$ can be considered as model mispecification.)

The solution to Eq. (18) is given by the stochastic integral

$$\begin{aligned} C(t) &= \exp\{-k(t-t_n)\}C(t_n) + [1 - \exp\{-k(t-t_n)\}]/(kV)\, r_n \\ &+ \int_{t_n}^{t} \exp\{-k(t-s)\}d\beta(s), \qquad t \in [t_n, t_{n+1}) \end{aligned} \tag{19}$$

To define a discrete time state equation corresponding to Eq. (1), set: $x_n = C(t_n)$, $\phi = (k, V)$ and

$$A(t, \tau, \phi) = exp\{-k(t - \tau)\} \tag{20}$$

$$B(t, \tau, \phi) = [1 - A(t, \tau, \phi)] / (kV) \tag{21}$$

$$W(t, \tau) = \int_{\tau}^{t} \exp\{-k(t - s)\} d\beta(s) \tag{22}$$

It follows from Eqs. (20)-(22):

$$x_{n+1} = A_n(\phi)x_n + B_n(\phi)r_n + W_n, \qquad n = 0, 1, \ldots, N \tag{23}$$

where $A_n(\phi) = A(t_{n+1}, t_n, \phi)$, $B_n(\phi) = B(t_{n+1}, t_n, \phi)$ and $W_n = W(t_{n+1}, t_n)$. It can be shown [1] that the sequence $\{W_n\}$ is independent Gaussian with mean 0 and variance

$$Q_n = Q \frac{\{1 - [B_n(\phi)]^2\}}{2k}$$

Additionally, if the infusion rate applied at time t_n is not exactly r_n but is equal to $r_n + \delta r_n$, where the error sequence $\{\delta r_n\}$ is independent Gaussian with mean 0 and variance R_n, then Eq. (23) becomes

$$x_{n+1} = A_n(\phi)x_n + B_n(\phi)r_n + w_n, \qquad n = 0, 1, \ldots, N \tag{24}$$

where the process noise

$$w_n = B_n(\phi)r_n + W_n$$

is independent Gaussian with mean 0 and variance $[B_n(\phi)]^2 R_n + Q_n$. Equation (24) is then the state equation corresponding to Eq. (1).

Measurement Equation

Assume noisy measurements $y(s_m)$ of the concentrations $C(s_m)$ are taken at the sampling times $\{s_m\}$ such that:

$$y(s_m) = C(s_m) + \epsilon_m, \quad m = 1, 2, \ldots, M$$

where ϵ_m is the assay noise of the mth measurement. The random variables $\{\epsilon_m\}$ are typically assumed to be independent Gaussian with mean 0 and standard deviation $\sigma_m = g(C(s_m))$. The function $g(C)$ comes from calibrating the assay device at various test concentrations C.

To define a discrete time measurement equation corresponding to Eq. (2), it is necessary to do some bookkeeping. Essentially, y_n is the collection of all measurements $\{y(s_m) : s_m \in [t_n, t_{n+1})\}$. The relationship between y_n and x_n is determined as follows. Let $S = \{s_1, \ldots, s_M\}$ be the set of sample times. Then for each $n = 0, 1, \ldots, N - 1$, either $S \cap [t_n, t_{n+1})$ is empty, or there are indices $m(n)$ and $m(n + 1)$ such that

$$S \cap [t_n, t_{n+1}) = \{s_{m(n)+1}, \ldots, s_{m(n+1)}\}. \quad (25)$$

Therefore, define:

$$\begin{aligned} y_n &= 0, && \text{if } S \cap [t_n, t_{n+1}) \text{ is empty.} \\ y_n &= (y(s_{m(n)+1}, \ldots, y(s_{m(n+1)})), && \text{if Eq. (25) holds.} \end{aligned} \quad (26)$$

(For example, if $N = 4$, $M = 3$, $s_1, s_2 \in [t_0, t_1)$ and $s_3 \in [t_3, t_4)$, then $m(0) = 0$, $m(1) = 2$, $m(3) = 2$, $m(4) = 3$ and $y_1 = (y(s_1), y(s_2))^T$, $y_2 = y_3 = 0, y_4 = y(s_3)$.)

Further, for $s_m \in [t_n, t_{n+1})$, it follows from Eqs. (19)-(22):

$$C(s_m) = A(s_m, t_n, \phi)x_n + B(s_m, t_n, \phi)r_n + W(s_m, t_n)$$

Therefore when Eq. (25) holds:

$$y_n = D_n(\phi)x_n + E_n(\phi)r_n + v_n \quad (27)$$

where

$$\begin{aligned} D_n(\phi) &= (A(s_{m(n)+1}, t_n, \phi), \ldots, A(s_{m(n+1)}, t_n, \phi))^T \\ E_n(\phi) &= (B(s_{m(n)+1}, t_n, \phi)), \ldots, B(s_{m(n+1)}, t_n, \phi)^T \\ v_n &= (W(s_{m(n)+1}, t_n), \ldots, W(s_{m(n+1)}, t_n)^T + (\epsilon_{m(n)+1}, \ldots, \epsilon_{m(n+1)})^T \end{aligned}$$

As before it can be shown that the sequence of random vectors $\{v_n\}$ is independent Gaussian with mean 0 and covariance depending on x_n and ϕ. Equations (26) and (27) are then the measurement equations corresponding to Eq. (2).

Optimal design of infusion regimens

In this case, the infusion times and sample times are fixed and the infusion rates are to be optimized. The controls are therefore given by $u_n = r_n$. The controls will be explicitly constrained to be nonnegative and implicitly constrained by the conditions:

$$E\{x_n\} \leq L_{max}, \quad n = 1, 2, \ldots, N+1 \quad (28)$$

where L_{max} is some maximum allowable level.

Cost Function: Target Level Control-Quadratic Cost

Here, we want $x_n \approx L_n$, where L_n is some desired target level at stage n. A suitable cost function corresponding to Eq. (7) is given by the quadratic cost function:

$$J(U) = E\{\sum_{n=0}^{N} \alpha_n(x_{n+1} - L_{n+1})^2 + \gamma_n(u_n)^2\} \quad (29)$$

where α_n and γ_n are given nonnegative constants reflecting the relative importance of the corresponding terms (including, for example $\gamma_n \equiv 0$). The optimization problem is to choose the admissible control policy to minimize Eq. (29).

Simulation Study: ***IPS*** **versus** ***SP***

In this simulation study we compare a separation principle policy with an iteration in policy space. The results presented here are taken from [14]. In the stochastic control formulation given by Eqs. (24), (27)–(29), the following specifications are made: $t_0 = 0, t_n = s_n = n, n = 1, \ldots, 5$. In this case Eqs. (24) and (27) become

$$\begin{aligned} x_{n+1} &= a x_n + b u_n + w_n, & x_0 = 0, \quad n = 0, \ldots, 5 \\ y_n &= x_n + v_n, & n = 1, \ldots, 5 \end{aligned}$$

where $a = \exp(-k)$ and $b = (1 - a)/(kV)$. A "natural" parameter vector for this model is then $\phi = (a, b)$.

For the cost function of Eq. (29) the targets and weights are chosen to be:

- $L_1 = 1.0$, $L_2 = 1.6$, $L_3 = 2.2$, $L_4 = 2.6$ and $L_5 = 2.8$.
- $\alpha_1 = \alpha_2 = \alpha_3 = \alpha_4 = 0.1$; $\alpha_5 = 1000$;
- $\gamma_0 = \gamma_1 = \gamma_2 = \gamma_3 = \gamma_4 = 0.0001$.

This choice of weighting indicates that we are essentially interested in attaining the desired level only at the fifth stage of the regimen. This is essentially a "terminal cost" problem and should reveal the importance of "active" learning. (Early mistakes are not severely penalized.) The controls u_n are nonnegative and the state constraint Eq. (28) is given by $E\{x_n\} \leq 3, \quad n = 1, 2, \ldots, 5$.

The prior distributions are chosen as follows: a, b, w_n and v_n are independent random variables such that:

$$\begin{aligned} w_n &\sim N(0, 0.01), \\ v_n &\sim N(0, 0.2), \\ a &\sim N(0.135, 0.0046) \quad \text{(C.V. = 34 \%)}, \\ b &\sim N(0.278, 0.0834) \quad \text{(C.V. = 30 \%)} \end{aligned}$$

where for any random variable v the notation $v \sim N(m, \sigma)$ means v is Gaussian with mean m and standard deviation σ.

Monte Carlo simulations of control performance were conducted. The simulations assessed comparative performance of an optimal separation principle (SP) type controller and an iteration in policy space (IPS) type controller. The SP controller employed an extended Kalman filter to calculate the conditional expectations of a, b, and x_n given the data I_n. The IPS controller was essentially a "wide-sense" approximation to one iteration in policy space using the previously calculated SP controller as the admissible nominal policy. (This particular algorithm is analogous to the method of Bar-Shalom and Tse [26]. See also [8,14] for more details.)

Figure 1 shows the control cost $J(U)$ in 100 simulations of the SP and IPS controllers. Not only is the mean cost for the IPS controller significantly lower than SP, but the standard deviation of the cost is also much lower. The existence of fewer outliers for IPS is readily apparent from the figure. The fact that the IPS performance is better than SP is expected from the definition of an IPS policy. The magnitude of this improvement was unexpected and it is argued that it comes from active learning.

Figure 2 shows the mean and one-sigma envelopes of the time histories of the serum concentrations for these same simulations. Note how the mean serum concentration of the IPS controller trajectory hugs the constraint boundary (exhibiting an active learning behavior). Also note that both controllers hit the target exactly "on average".

EXAMPLE 2: OPTIMAL SAMPLING SCHEDULE DESIGN

In this section we consider the classical PK/PD problem of parameter estimation and sampling schedule design. Assume noisy measurements $y(s_n)$ are taken at the sampling times $\{s_n\}$ from a PK/PD system modeled by a "nonlinear regression" equation of the form:

$$y(s_n) = h_n(s_n, \phi) + \epsilon_n, \quad n = 1, 2, \ldots, N \tag{30}$$

where $h_n(s, \phi)$ is a known function and ϵ_n is the error of the nth measurement. In Eq. (30) the following assumptions are made: The random variables $\{\epsilon_n\}$ are independent Gaussian with mean 0 and standard deviation σ_n, which is independent of ϕ but may depend on s_n. The function $h_n(s, \phi)$ has continuous partial derivatives

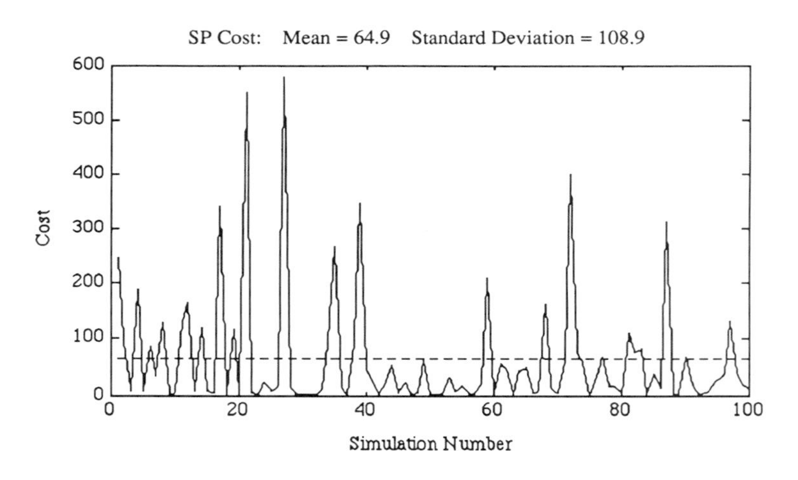

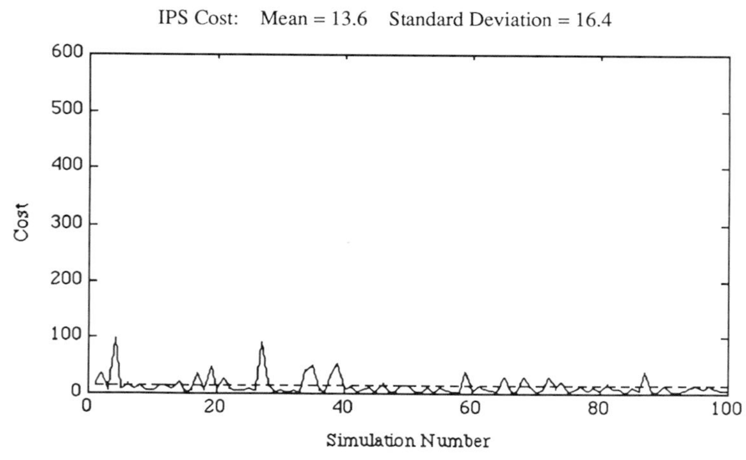

Fig. 1. *SP* and *IPS* costs.

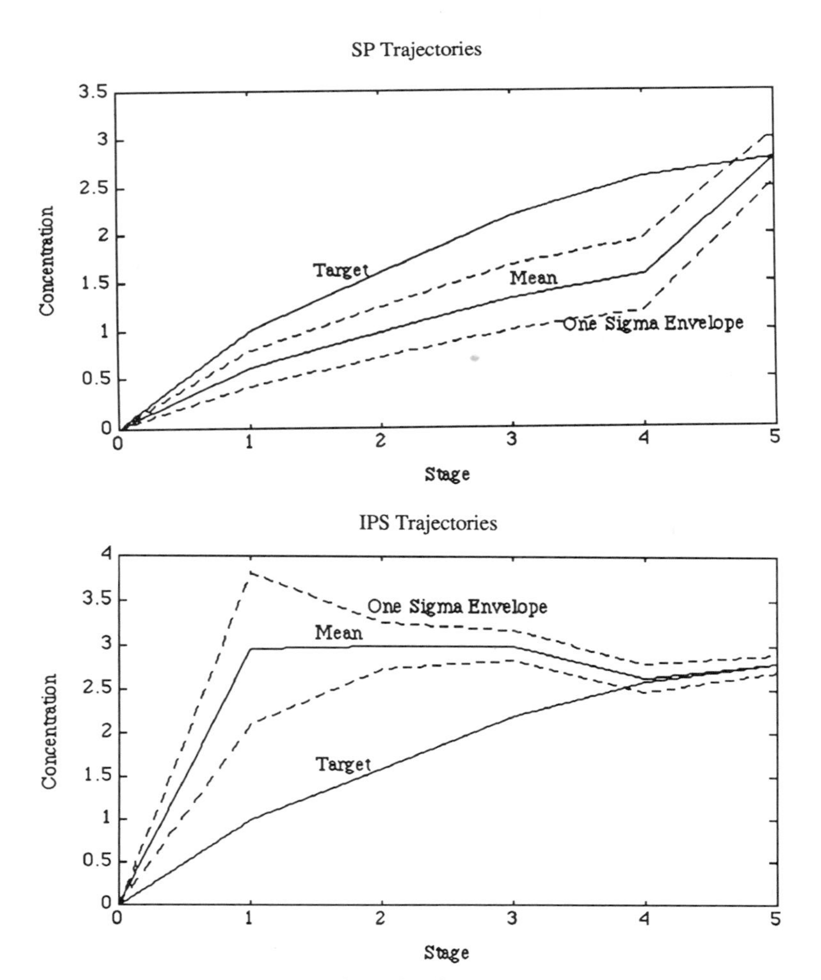

Fig. 2. *SP* and *IPS* trajectories.

with respect to the components of ϕ. The PK/PD vector ϕ is random with known probability density $p(\phi)$.

D-optimality and variants

The problem here is to design a sampling schedule so that the resulting estimate of ϕ provides as much "information" as possible. The most common measure of this information is the so-called "D-optimality" criterion; that is the criterion given in Eq. (8). Similar but not identical criteria occur when the det(M) in Eq. (9) is replaced by log det(M). For the model of Eq. (30), it can be shown that the Fisher information matrix is:

$$M(\phi, U) = \sum_{n=0}^{N} \eta_n(\phi, s_n)^T \eta_n(\phi, s_n)/\sigma_n^2, \tag{31}$$

where $\eta_n(\phi, s) = \partial h_n(s, \phi)/\partial\phi = (\partial h_n(s, \phi)/\partial\phi_1, \ldots, \partial h_n(s, \phi)/\partial\phi_p)$ and $\phi = (\phi_1, \ldots, \phi_p)^T$.

In general, the objective is to maximize $E\{\Phi[M(\phi, U)]\}$ for some suitable function Φ. We now put this problem into the stochastic control framework developed above.

First define the "controls" $u_n = s_{n+1}$ with control constraints: $0 \leq u_0 \leq u_1 \leq \ldots \leq u_N \leq T$. The measurement equation corresponding to Eq. (2) is obtained by setting $y_n = y(s_n)$ and $v_n = \epsilon_n$ in Eq. (30):

$$y_n = h_n(u_n, \phi) + v_n \tag{32}$$

(Note that there are no "state" variables in this equation.)

The main purpose of the "state" equation is to get the cost function in the form of Eq. (10). To this end, define the "matrix" state variables $\{x_n\}$ by the state equation:

$$x_{n+1} = x_n + \eta(\phi, u_n)^T \eta(\phi, u_n), \qquad x_0 = 0 \qquad n = 0, 1, \ldots, N$$

(Note that the state equation is deterministic given ϕ.) If we define:

$$C(X, U, \phi) = -\Phi(x_{N+1})$$

then the resulting optimization problem becomes one of minimizing a "terminal cost":

$$J(U) = -\int \Phi(x_{N+1}) p(\phi) d\phi \tag{33}$$

over all admissible controls U.

In the PK/PD literature, the D-optimality problem (and its variants) have not been considered explicitly from such a stochastic control framework. However a number of previous works can be interpreted in this light. For example, in D'Argenio [27] and Walter and Pronzato [28], optimal "open loop" controllers are obtained; and in D'Argenio [29] an optimal "separation principle" controller is derived. (See also [30] for a recent survey on this subject.)

Considered as a terminal cost problem, the optimal sampling schedule design should be ideally suited for optimal policies. In [31] an iteration in policy space approach is suggested. A "natural" nominal admissible control policy is available. Namely:

$$U_0 \equiv (\mu_0, \ldots, \mu_N)$$

is such that, at each stage n, the remaining points $\{\mu_i, i \geq n\}$ are equally distributed in the open interval (μ_{n-1}, T), where $\mu_{-1} \equiv 0$.

It is hoped that this control perspective brings new insight to optimal sampling design.

ACKNOWLEDGMENTS

The writing of the paper benefited in many ways from many people; and the author wishes to express his sincere appreciation to David S. Bayard, David Z. D'Argenio, Marianne Hubner, Roger W. Jelliffe, Mark Milman and Poornima Raghu. Further, this research was supported in part by National Institutes of Health grants RR01629 and P41-RR01861.

REFERENCES

1. A. .H. Jazwinski. *Stochastic Processes and Filtering Theory*, Academic Press, New York, 1970.
2. A. Racine-Poon and A. F. M. Smith. Population models. In D. A. Berry and Marcel Dekker (eds.), *Statistical Methodology in the Pharmaceutical Sciences*, New York, 1990, pp. 139–162.
3. J. L. Steimer, A. Mallet, and F. Mentré. Estimating interindividual pharmacokinetic variability. In M. Rowland *et al.* (eds.), *Variability in Drug Therapy: Description, Estimation, and Control*, Raven Press, New York, 1985, pp. 65–111.
4. R. W. Jelliffe. A simulation study of factors affecting aminoglycoside therapeutic precision. In C. Cobelli and L. Mariani (eds.), *Proc. First Symposium on Modeling and Control in Biomedical Systems*, Venice, Italy, 1988, pp. 86–88.
5. R. W. Jelliffe, A. Schumitzky, and M. Van Guilder. A simulation study of factors affecting aminoglycoside therapeutic precision, Technical Report: 90-3, Laboratory of Applied Pharmacokinetics, USC School of Medicine, Los Angeles, 1990.
6. D. P. Bertsekas. *Dynamic Programming: Deterministic and Stochastic Models*, Prentice Hall, Englewood Cliffs, 1987.
7. R. Bellman. *Adaptive Control Processes: A Guided Tour*, Princeton University Press, Princeton, 1961.
8. D. S. Bayard. A forward method for optimal stochastic nonlinear and adaptive control. In *Proc. 27th IEEE Conference on Decision and Control*, Austin, 1988. To appear: *IEEE T. Automat. Contr.*
9. D. S. Bayard. Aspects of stochastic adaptive control synthesis. Doctoral Thesis, Electrical Engineering Department, State University of New York, Stony Brook, 1984.
10. Y. Bar-Shalom. Stochastic dynamic programming: caution and probing. *IEEE T. Automat. Contr.* **AC-10**:1184–1195 (1981).
11. D. S. Bayard and M. Eslami. Implicit dual control for general stochastic systems. *Opt. Cont. Appl. Meth.* **6**:265–279 (1985).
12. W. F. Powers, P. H. Abbrecht, and D. G. Covel. Systems and microcomputer approach to anticoagulant therapy. *IEEE T. Bio-med. Eng.* **27**:520–523 (1980).
13. S. Amrani, E. Walter, Y. Lecourtier, and R. Gomeni. Robust control of uncertain pharmacokinetic models. *Proc. IFAC 9th Triennial World Congress, Budapest*, 3079–3083 (1984).
14. A. Schumitzky, M. Milman, P. Khademi, and R. Jelliffe. Approximate optimal closed loop control of pharmacokinetic systems. In *Proc. IFAC Workshop on Decision Support for Patient Management: Measurement, Modeling, and Control*, British Medical Informatics Society, London, 1989, pp. 338–347.
15. J. Gaillot, J-L. Steimer, A. Mallet, J. Thebault, and A. Beider. A prior lithium dosage regimen using population characteristics of pharmacokinetic parameters. *J. Pharmacokin. Biopharm.* **7**:579–628 (1979).
16. O. Richter and D. Reinhardt. Methods for evaluating optimal dosage regimens and their application to theophylline. *Int. J. Clin. Pharm. Th.* **20**:564–575 (1982).
17. A. Mallet, F. Mentré, J. Giles, A. W. Kelman, A. H. Thomson, S. M. Bryson, and B. Whiting. Handling covariates in population pharmacokinetics with an application to gentamicin. *Biomed. Meas. Infor. Contr.* **2**:138–146 (1988).
18. D. Katz and D. Z. D'Argenio. Stochastic control of pharmacokinetic systems: Open loop strategies. In C. Cobelli and L. Mariani (eds.), *Proc. First Symposium on Modelling and Control in Biomedical Systems*, Venice, 1988, pp. 560–566.
19. D. Z. D'Argenio and D. Katz. Implementation and evaluation of control strategies for individualizing dosage regimens with application to the aminoglycoside antibiotics. *J. Pharmacokin. Biopharm.* **14**:523–37 (1986).
20. L. B. Sheiner, B. Rosenberg, and K. L. Melmon. Modelling of individual pharmacokinetics for computer-aided drug dosage. *Comput. Biomed. Res.* **5**:441–459 (1972).
21. R. W. Jelliffe. Clinical applications of pharmacokinetics and control theory: planning, monitoring, and adjusting dosage regimens of aminoglycosides, lidocaine, digitoxin, and digoxin. In R. Maronde (ed.), *Selected Topics in Clinical Pharmacology*, Springer-Verlag, New York, 1986, pp. 26–82.

22. R. W. Jelliffe, A. Schumitzky, and L. Hu, M. Liu. PC computer programs for Bayesian adaptive control of drug dosage regimens. Technical Report: 90-5, Laboratory of Applied Pharmacokinetics, USC School of Medicine, Los Angeles, 1990.

23. S. Vozeh and J. L. Steimer. Feedback control methods for drug dosage optimization. *Clin. Pharmacokinet.* **10**:457–476 (1985).

24. A. Schumitzky. Stochastic control of pharmacokinetic systems. In R. Maronde (ed.), *Selected Topics in Clinical Pharmacology and Therapeutics*, Springer-Verlag, New York, 1986, pp. 13–25.

25. A. Schumitzky. Adaptive control in drug therapy. In H. Ducrot *et al.* (eds.) *Computer Aid to Drug Therapy and to Drug Monitoring*, Berne, Switzerland, March 6-10, 1978, North Holland, Amsterdam, 1978, pp. 357–360.

26. Y. Bar-Shalom and E. Tse. Concepts and methods in stochastic control. In C. Leondes (ed.), *Control and Dynamic Systems*, Vol. 12, Academic Press, New York, 1976, pp. 99–172.

27. D. Z. D'Argenio. Incorporating prior parameter uncertainty in the design of sampling schedules for pharmacokinetic parameter estimation experiments. *Math. Biosci.* **99**:105–118 (1990).

28. E. Walter and L. Pronzato. Robust experimental design via stochastic approximation. *Math. Biosci.* **75**:103–120 (1985).

29. D. Z. D'Argenio. Optimal sampling times for pharmacokinetic experiments. *J. Pharmacokin. Biopharm.* **9**:739–56 (1981).

30. L. Pronzato and E. Walter. Qualitative and quantitative experiment design for phenomenological models - A survey. *Automatica* **26**:195–213 (1990).

31. D. S. Bayard and A. Schumitzky. A stochastic control approach to optimal sampling schedule design. Technical Report: 90-1, Laboratory of Applied Pharmacokinetics, USC School of Medicine, Los Angeles, 1990.

PHARMACOTHERAPEUTICS: MEASUREMENT, CONTROL AND DELIVERY

DEVELOPMENT OF A FIBER OPTIC SENSOR FOR DETECTION OF GENERAL ANESTHETICS AND OTHER SMALL ORGANIC MOLECULES

Sabina Merlo and **Paul Yager**

Center for Bioengineering
University of Washington
and
The Washington Technology Center

Lloyd W. Burgess

Center for Process Analytical Chemistry
University of Washington

ABSTRACT

We have demonstrated a fiber optic sensor for general anesthetics and other small organic molecules based on the sensitivity of the phase transitions of phospholipids to the presence of such molecules. An aqueous dispersion of phospholipids that exhibit a phase transition near the desired operating temperature is used as a transducing element at the end of two optical fibers. Agarose gel is used to immobilize liposomes labeled with a fluorescent dye sensitive to the phase state of the bilayer in which it is embedded. The ratio of the fluorescence intensity at two wavelengths exhibits a nearly linear dependance on clinically relevant concentrations of volatile anesthetics over a physiological range of temperatures. The probe has been tested in both gas and liquid phases, and should operate in blood and tissue, allowing development of a practical device for rapid *in vivo* monitoring of general anesthetics.

INTRODUCTION

An important objective of the monitoring of patients during anesthesia and surgery is to assess depth of anesthesia. It is essential to balance avoidance of the patient's awareness of the surgical environment with minimization of the risk associated with excessive doses of anesthetic drugs. The effectiveness of anesthesia may be evaluated by monitoring muscle and hemodynamic responses to stimuli, central nervous system activity (through EEG and evoked potentials), pattern of ventilation, and pupillary size. However, these measurements are not necessarily specific for anesthesia because the monitored parameters may be altered by other factors, such as use of muscle relaxants and changes in the physiological status of the patient [1].

Alternatively, the concentration of anesthetic drugs in gas or blood samples can be measured directly using techniques such as gas chromatography, mass spectroscopy, and IR or UV gas analyzers. The concentrations of inhaled anesthetics are measured as partial pressure in the inspired gas. An index of anesthetic equipotency called the MAC (minimum alveolar concentration) has been established that,

Advanced Methods of Pharmacokinetic and Pharmacodynamic Systems Analysis
Edited by D'Argenio, Plenum Press, New York, 1991

in humans, is empirically defined as the alveolar anesthetic concentration at which 50% of patients respond by movement to surgical incision [1]. The anesthetic partial pressure that corresponds to 1 MAC (in % of 1 atm) for three commonly used volatile anesthetics is 1.15% for isoflurane, 1.68% for enflurane, and 0.7% for halothane. For an inert anesthetic the alveolar partial pressure should, at equilibrium, equal that in the central nervous system. This equilibrium value should be independent of the total uptake and distribution of the anesthetic agent to different tissues. However, if the anesthetic is metabolized or excreted at an appreciable rate, then a dynamic equilibrium is established and these assumptions do not hold. Also, during induction and emergence from anesthesia, the concentration of the anesthetic in various tissues will differ from that in the lungs.

A drawback of some techniques used to monitor anesthetic concentration is that some of the instrumentation may not be located near the patient, so samples have to be transferred to an analytical laboratory, resulting in long response times and possible sample degradation. Clearly, it would be advantageous to have a sensor that allowed continuous, *in vivo* monitoring of general anesthetics. While there is no such device currently available, fiber optic sensors have been investigated for this purpose. Wolfbeis et al. have demonstrated a fiber optic sensor for remote detection of halothane based on dynamic quenching of the fluorescent dye decacyclene incorporated into a silicone rubber membrane [2]. The bromine atom on halothane is responsible for quenching of the dye; other common general anesthetics are not brominated and are not detectable. A fiber optic probe employing remote infrared absorption spectroscopy for monitoring the uptake of anesthetic gases in patients during surgery has also been described [3]. In related work, Posch et al. have developed a fiber optic sensor for vapors of polar organic solvents based on reversible decolorization of blue thermal printer paper [4], however, their optical method does not respond to common anesthetic agents, such as isoflurane and enflurane. None of these probes has been demonstrated to operate in blood or tissue.

We recently demonstrated a convenient fluorescence method for monitoring changes in *fluidity* of labeled liposomes, and showed that it could be applied to fluidity changes caused by absorption into the lipids of clinical concentrations of general anesthetics [5]. Changes in lipid order produced substantial fluorescence spectral shifts, which could be quantified by taking the ratio of the fluorescence intensity at two wavelengths. Monitoring the order of a lipid bilayer system in the presence of anesthetics relative to the order at the same temperature in absence of anesthetic allows determination of the anesthetic concentration in the bilayer. Bilayer concentration, in turn, is linearly proportional to anesthetic concentration in the environment around a sample of lipids.

What follows is a demonstration that the previously described lipid system and optical method can be applied for the remote detection of anesthetic concentration by combining the physical chemistry of lipids with fiber optic sensor technology. The sensor does not consume analyte, so the equilibrium sensor response should not be dependent on biofouling, and its response time is limited only by diffusion of small molecules into the probe. It exhibits a nearly linear response to clinically relevant concentrations of volatile anesthetic in the gas phase over a physiological range of temperatures, and should operate equally well in liquids such as blood and cerebrospinal fluid. The described device is not specific for a particular anesthetic, but responds to hydrocarbons, alcohols, and other lipid-soluble chemicals in the sample matrix. The lack of selectivity of the transducing element could be an advantage in clinical situations where the combined effect of several anesthetic drugs must be monitored. A preliminary version of this work has been published elsewhere [6].

EXPERIMENTAL SECTION

Reagents

1,2-bis(palmitoyl)-**sn**-glycero-3-phosphocholine (DPPC) and 1,2-bis(myristoyl)-**sn**-glycero-3-phosphocholine (DMPC) were purchased in chloroform solution (purity >99%) from Avanti Polar Lipids Inc., and used as obtained. The fluorescent dye 6-lauroyl-2-(dimethylamino)-napthalene (Laurdan) was obtained from Molecular Probes. Agarose Type 1-A No. A-0169 was purchased from Sigma Chemical. Three commonly used general anesthetics, isoflurane (Forane®; 1-chloro-2,2,2-trifluoroethyldifluoro-methyl ether), enflurane (Ethrane®; 2-chloro-1-(difluoromethoxy)-1,1,2-trifluoroethane) and halothane (Fluothane®; 2-bromo-2-chloro-1,1,1-trifluoroethane), were donated by the Anesthesiology Department of the University of Washington. Ethanol (100%) was distilled prior to use.

Fluorescent-labeled hydrated lipid was prepared in a manner similar to that previously described [5] as follows. Laurdan and one or more phospholipids (molar ratio 1:150) were first codissolved in chloroform. The solvent was evaporated in an Evapotec Micro Rotary Film Evaporator (Haake Buchler Instruments, Inc.) while the container was warmed in a water bath at 50°C. In order to eliminate any trace of solvent, the sample was then stored under vacuum for 8 hours. Dried lipid was then rehydrated with buffer (10 mM HEPES / 100 mM NaCl in distilled deionized water, pH 7.5). The samples were usually stored at 4°C overnight. Liposomal dispersions of vesicles were prepared the next day by vortexing the hydrated lipids using a Vortex Genie Mixer (Scientific Products) at setting 7 for a total of 10 min in bursts of 1 min interspersed with reheating to above the phase transition temperature [5]. Agarose gel was prepared by hydrating agarose powder with aqueous buffer. The aqueous agarose solution was then held above its gelling temperature for 10 min. Next, the suspension of lipid vesicles in buffer, which had been kept above the phase transition temperature after vortexing, was mixed into the warm agarose at the desired concentration and was immediately used for immobilization in the optrode as described below. Best results were obtained with a final lipid concentration of 75 mg/ml in a 1% agarose gel.

Instrumentation

Source: The fluorescence excitation source was a short arc mercury vapor lamp (Osram HBO 50 Watt Super Pressure mercury lamp) with an 11 nm bandwidth band pass filter centered at 365 nm (Carl Zeiss, Inc). The excitation light was coupled into an optical fiber using a 25X Planachromat microscope objective lens (N.A.=0.45)(Carl Zeiss, Inc). This fiber was placed in the fiber-holder of an x-y positioner that was part of the microscope stage used for optical alignment.

Spectrometer: The backscattered radiation at 365 nm and emitted fluorescence were collected by a second optical fiber directly coupled to a Chemspec 100S spectrometer (f/2.2) (American Holographic, Inc.). The spectrum was dispersed and imaged on the detector by a concave aberration-corrected holographic diffraction grating with a dispersion of 20 nm/mm operating in the range 300–800 nm.

Detector: The detector was an EG&G Reticon "S" series self-scanning linear photodiode array (RL0512S). It consisted of 512 diodes and was 12.8 mm long. We were able to image on the detector only the spectral region corresponding to fluorescence emission and thus, remove most of the 365 nm backscattered light. Further attenuation of stray light at the detector was obtained by placing a high pass filter

(Melles Griot sharp cut-off, 03 FCG 055, Schott glass type GG 395, 50% transmission at 395 nm) in front of the detector. All the signal and video processing circuitry necessary to operate the photodiode array were provided by the EG&G boards (RC1000 mother board and RC1001 satellite board).

Data acquisition: The system for data acquisition and storage was comprised of an expansion board (DASH-16 from MetraByte Corporation) internally installed in an IBM/XT Personal Computer. The analog-to-digital converter (ADC) was triggered by a direct external trigger to the DASH-16 and data were transferred by Direct Memory Access (DMA) to obtain high data throughput. In our software, a machine language driver (DASH16.BIN) controlled A/D conversion and digital I/O channel functions, and data transfer was via BASIC CALL. Initial setup and installation aids were also included. Additional Basic software was written for subtraction of the background from the collected spectra and for calculation of the intensity ratio. This system required 400 msec for collection of each full spectrum, although to produce the ratio R additional time was required for data manipulation.

Optrode Design

Of several optrode designs attempted the most successful approach utilized an 18 gauge hypodermic needle with a side hole as a support for fibers and transducing chemistry. A drawing of the needle optrode is shown in Fig. 1. The optical fibers employed for this configuration were HCN/H high N.A. (N.A.=0.44–0.48) HCS®multimode optical fibers (Ensign-Bickford Optics Company). A 200 μm core diameter optical fiber carried the excitation light to the probe tip, while a 125 μm core diameter fiber collected the fluorescence signal. Removal of the mechanical buffer coating from the sensing end of the larger fiber allowed closer approach of the parallel ends of the fibers and thus, increased collection efficiency. The fibers were manually cleaved, tips aligned, glued together, and fixed in the needle so that the endfaces were at the level of the side hole. A drop of warm fluid agarose-liposome mixture was then deposited through the side hole and allowed to gel around the fibers on cooling. The volume was sufficient to overfill the cavity, thereby providing good adhesion to both fibers and needle. Dehydration of the gel was prevented by storing the probe over water in a sealed vial. No difficulty was encountered in using

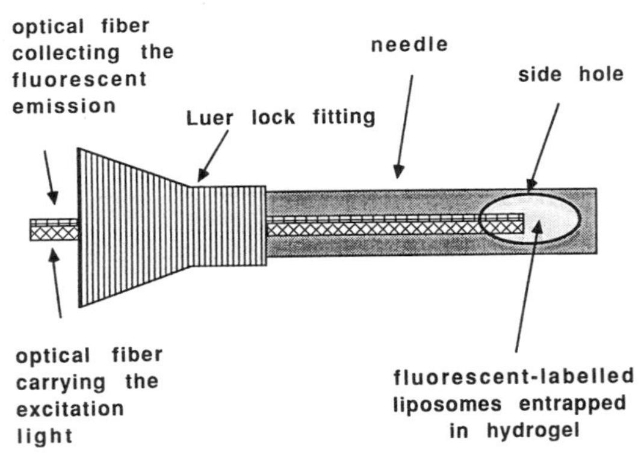

Fig. 1. Design of the optrode using a modified 18 gauge stainless steel hypodermic needle as a rigid support. The tip was ground flat shut and a 3 mm long side hole was introduced to allow access to the fiber tips. Note that the hydrogel generally overfilled the cavity surrounding the fibers.

lipid concentrations as high as 75 mg/ml in the probe, and as increased liposome concentrations produced stronger fluorescent signals, this concentration was chosen for testing the response of the probe to anesthetic.

We were able to utilize high N.A. fibers for efficient light excitation and collection in the optrode. The high N.A. collection fiber did not significantly overfill the spectrometer's aperture because the effective acceptance angle of the fiber in the optrode tip is lowered by the high refractive index of the lipid-gel suspension. An underestimated value for this refractive index is 1.4 [7], which would produce an entrance aperture ranging from 0.31–0.34 in the probe. Redistribution of the collected fluorescent light into the full fiber aperture does not occur due to the short length of the fiber employed.

Experimental procedure

Initially, the response of the probe to temperature was determined in the absence of anesthetic. These experiments were performed by placing the optrode into a sealed 1087 ml flask that contained 30 ml of deionized water to prevent dehydration of the gel in the probe. The flask sat in a thermostatically controlled water bath (Exacal EX300, Neslab Instruments Inc.). The temperature was monitored with a Teflon-coated thermocouple placed beside the optrode and connected to a Sensortek BAT-10 digital thermometer with a resolution of 0.1°C. The optical fibers and the thermocouple were passed through the Teflon septum of an open-top screw cap that was used for closing the flask. Temperature was stable to within ±0.1°C.

We followed a procedure described in our previous publication on this lipid system [5] to determine the two emission wavelengths that undergo the largest change in value for a given change in lipid order. Fluorescence emission spectra were initially collected with a probe made of pure DPPC vesicles entrapped in 1% agarose gel. The difference between spectra collected at temperatures immediately below and above this sample's gel-liquid crystal phase transition at 41.4°C showed that the largest changes in intensity occurred at 445 and 517 nm. For subsequent measurements, an intensity ratio R was calculated by subtracting from a full emission spectrum the background due to the dark current of the diode array, summing the readouts of diodes 139 to 144 (corresponding to 444–446 nm), and dividing the resultant number by the sum of the intensities on diodes 280 to 285 (corresponding to 516–518 nm). We have noted that a ratio generated in such a manner has the greatest possible sensitivity to changes in order, but does not vary linearly with the mole fraction of gel phase in the sample [5]. However, R can be used in a reproducible, if semiquantitative manner to monitor the degree of conversion from gel to liquid crystal state in the vicinity of the phase transition. Note that both the wavelengths used and R values found at a given temperature differ slightly from those derived from previous experiments using a Perkin-Elmer spectrofluorimeter [5] because of differences in optical characteristics of the two systems.

Varying concentrations of the anesthetics isoflurane, enflurane, and halothane in the gas phase were then monitored at constant temperature. Experiments were performed using the same equipment by injecting a known volume of liquid anesthetic into the sealed flask using a gas-tight syringe, and allowing it to come to equilibrium [8]. The temperature dropped a few tenths of a degree after injection, but returned to the starting value in 2–3 min. Different concentrations of anesthetic in the gas phase in the absence of the probe were verified using a Varian Aerograph series 1200 gas chromatograph with a flame ionization detector, and were found to be equal (±0.05% of 1 atm) to the values calculated with ideal gas law in a dry empty flask.

These theoretical values are reported in the figures.

The response of the probe to ethanol in water was also studied at constant temperature. For this experiment the needle was inserted through a Teflon septum before assembling the optrode. The probe was then placed into a Pierce Reacti-Vial^{TM} reaction vial with a screw cap incorporating a Teflon septum. The vial contained 5.0 ml of deionized water. Different ethanol concentrations were created by injecting ethanol through the septum with a gastight syringe. Since each addition of ethanol increased the total liquid volume, we defined the concentration in mg of ethanol per ml of water. Each addition consisted of 16 μl of ethanol, corresponding to 2.5 mg of ethanol per ml of water.

RESULTS AND DISCUSSION

Response to temperature

Fluorescence spectra were collected with probes made of different ratios of DPPC to DMPC over a range of temperatures. Fluorescence spectra collected during a heating scan between 41.1°C and 45.7°C from a probe made of pure DPPC vesicles in air are shown in Fig. 2. These spectra are qualitatively similar to those recorded from liposome samples in cuvettes using a conventional fluorimeter [5], with the exception that in this case a signal-to-noise ratio (SNR) of nearly 100 was obtained with each single 400 msec scan, even after background subtraction. The intensity ratio R plotted as a function of temperature clearly showing the change in bilayer order at T_m is illustrated in Fig. 3. Note that the isosbestic point for the emission was identifiable only if spectra were rapidly collected in a small temperature range. This was a consequence of the decrease in absolute fluorescence intensity during the course of the experiment caused by photobleaching of the dye. Photobleaching was not found to be a problem in previous experiments with a fluorimeter [5]; in those experiments with liposomes prepared in a similar manner an isosbestic point was consistently found. Since the photobleaching appeared to decrement the entire emission spectrum uniformly, the value of R was not affected, even when the intensity

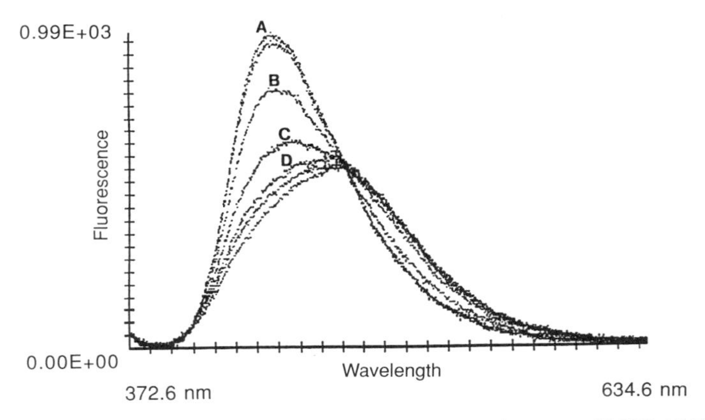

Fig. 2. Fluorescence spectra collected during a heating scan in the range 41.1°C-45.7°C with a pure DPPC probe. Each curve represents a single 400 msec collection period from which background spectra collected with the excitation lamp blocked have been subtracted. The labeled emission spectra were recorded at the following temperatures (A) 41.1°C, (B) 41.7°C, (C) 41.9°C, (D) 42.5°C.

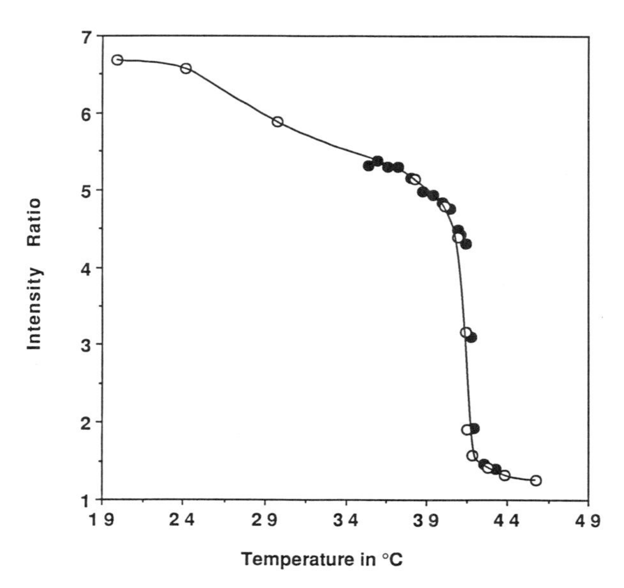

Fig. 3. Response of the intensity ratio R from a pure DPPC probe to temperature changes: ● heating scan; ○ cooling scan. Closer inspection of the two curves revealed a 0.3°C shift between the cooling and heating curves near the phase transition at 41.4°C. The solid line is a polynomial fit to the cooling data.

dropped by more than 50%. Photobleaching of the dye did limit the lifetime of the probes, as determined by how long a usable SNR was maintained. This problem was substantially reduced by manually opening a shutter in front of the source for intervals of a few seconds during which spectra were collected. Probes remained usable after as long as 15 minutes of continuous illumination. Probe lifetime could be substantially extended by improvement of the manual shuttering method.

With the present optrode design the use of pure DPPC vesicles precludes operation near 37°C. We have shown that the Laurdan fluorescent probe monitors the broadening and shifting of the DPPC phase transition when increasing proportions of DMPC are added [5]. The simple technique of admixture of a *contaminant* can be used to adjust the temperature of maximum sensitivity and the breadth of the operating range in a *phase transition sensor*. Mixed lipid systems were also effective for tailoring the thermal response of the sensor probes used in this study, allowing the probe to be sensitive over several degrees bracketing physiological temperatures.

Hysteresis was occasionally observed between heating and cooling cycles in some probes; that is, the same value of R was not always immediately obtained when a probe approached the same temperature from above or below. Such hysteresis could produce unreliable sensor operation if the probe were heated to temperatures above its anticipated operating range, or if at operating temperatures the probe were exposed to anesthetic concentrations far in excess of those anticipated. Possible causes for the sporadically observed hysteresis include the following: a) variation in the placement of the thermocouple relative to the hydrogel from probe to probe, b) history-dependent alterations in liposome response due to dehydration or changes in the gel-liposome interaction, or c) intrinsic hysteresis of the lipid phase transition.

The intrinsic relaxation time within the phase transition for large DPPC or DMPC liposomes of this sort is no longer than a few seconds in either heating or cooling directions [9], and for small liposomes the time is even shorter. Hysteresis is not an intrinsic characteristic of the main phase transitions of such liposomes. The measured T_m can be affected by the rate of change of the temperature; for example, the shape of the calorimetrically observed phase transition of DPPC/DMPC mixtures as measured in heating scans is similar but not identical to that observed on cooling, even at 0.2°C per min [10]. At finite rates of cooling, formation of crystallization nuclei may be rate limiting in the freezing process, resulting in possible supercooling. However, if the material is never completely melted and some crystalline material remains, rapid re-freezing can occur and there should be no hysteresis. No correlation between cooling rates and observed hysteresis was evident, but hysteresis was consistently found on cooling immediately after large excursions above T_m, whereas when the temperature exceeded T_m by no more than a few degrees, no hysteresis whatsoever was found. This suggests that cause c is responsible, but is also easily avoidable. Reversible sensor operation should be possible within the phase transition region itself; such can easily be the case when the sensor uses a DPPC/DMPC mixture with a broad phase transition. "Resetting" an accidentally overheated sensor might be accomplished by briefly cooling to temperatures below the phase transition, thereby regenerating crystallization nuclei. In these initial experiments the precaution of remaining within the phase transition was not always taken. We are currently attempting to confirm this hypothesis of the origin of the intermittent hysteresis observed.

Response to general anesthetics

It is well-known that partitioning of general anesthetics into lipid bilayers results in a depression of the phase transition of lipid bilayers that is linearly proportional to anesthetic concentration [11]. The new method for detecting general anesthetics consists of comparing the fluorescence intensity ratio R from Laurdan-labeled phospholipid vesicles at a defined temperature in the absence and presence of anesthetic. This method was not sufficiently sensitive to detect the presence of anesthetic if the temperature was much lower or higher than the phase transition temperature T_m of the lipid. However, the method was very sensitive if the measurement was performed at a temperature just below T_m. The sharper the phase transition, the greater the sensitivity to anesthetic, but the narrower is the range of temperatures over which there is such sensitivity.

The first tests were of the response to anesthetics of a probe made of pure DPPC vortexed vesicles immobilized in agarose gel. In this and subsequent experiments a probe was warmed in the closed humidified flask until R dropped to about 4 near the onset of melting. For pure DPPC vortexed vesicles this occurred at 41°C. Six 25 μl aliquots of isoflurane were added to the flask and R was measured after a period of equilibration following each injection. Figure 4 allows a comparison of the values of R observed after 15 and 30 min. As expected, the sigmoid shape of the data mirrors the response to temperature observed in previous experiments. The difference between 15 and 30 min data implies that a slow equilibration process is occurring in the system. Similar results were noted in the following experiments. There are two probable contributions to the response time. The liquid anesthetic volatilizes quickly, but the equilibration of concentration in the closed system may be slower and may be affected by the water in the flask. A second factor is the diffusion time of the anesthetic into the lipid-gel phase contained in the probe. Several microliters of lipid suspension surrounded the optical fibers–a much larger volume than that contributing to the fluorescence signal. The anesthetics had to diffuse through at least 1 mm of rigid hydrogel to reach the optically sampled volume (and

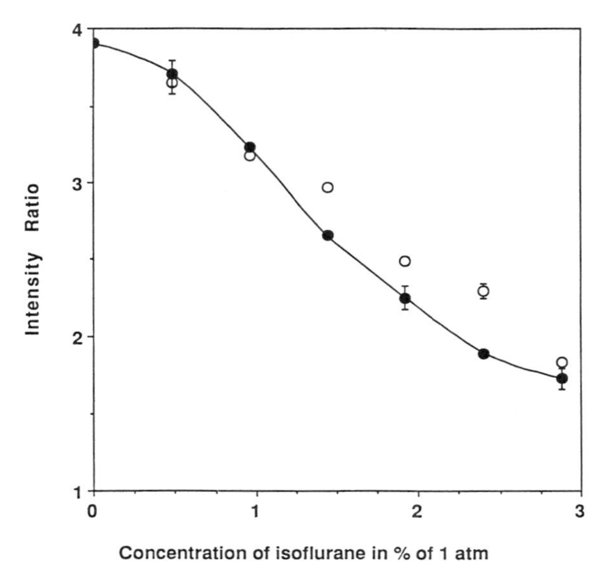

Fig. 4. Effect of equilibration time on the values of R as measured with a pure DPPC probe. Data were collected: ○ after 15 min.; ● after 30 min.

had to equilibrate into relatively inaccessible dead-end compartments within the steel needle as well) before equilibration was reached. It takes approximately 10 minutes for the somewhat smaller propane molecule to diffuse through 1 mm of water at 25°C. A 1% hydrogel is very similar to unstirred water, so the observed response times are completely in accord with theory. Reduction of the diffusion path is necessary to achieve more rapid response. A 10 second response time, for example, requires a diffusion path of no more than 100 μm.

The response to the anesthetic enflurane was determined using a probe made of a DPPC/DMPC lipid mixture in the molar ratio 90:10. We have shown that in such mixtures R varies approximately linearly with temperature over a nearly 2°C range [5]. The experiment was performed at 38.5°C and the results are shown in Fig. 5, where the intensity ratio is plotted as a function of the theoretical partial pressure of enflurane generated in the flask by adding eight 25 μl aliquots of liquid enflurane. In addition to a temperature of operation lower than that with pure DPPC, this probe is characterized by a nearly linear relationship between R and anesthetic concentration. Probes made of a DPPC/DMPC lipid mixture in the molar ratio 86:14 (which has an even broader linear response to temperature than does 90:10) were also tested in response to isoflurane at 38.2°C and enflurane at 38.1°C. The results for isoflurane collected after 15 and 30 min are presented in Fig. 6. Good linearity of the intensity ratio versus anesthetic partial pressure is observed, in particular for the 30 min data. With no signal averaging it was possible to resolve 0.2 units of intensity ratio, which corresponds to an error in the anesthetic concentration of 0.4% (a slope of approximately -0.5 was obtained in this case). In a clinical situation an error of 0.2% can be accepted and commercially available gas phase monitors are able to resolve 0.1%. The very similar results for enflurane collected after 15 and 30 min are presented in Fig. 7. Linearity of the intensity ratio R *versus* anesthetic partial pressure was again observed. The response to isoflurane of a probe made of a DPPC/DMPC mixture in 80:20 molar ratio was also tested to demonstrate the possibility of detecting clinically

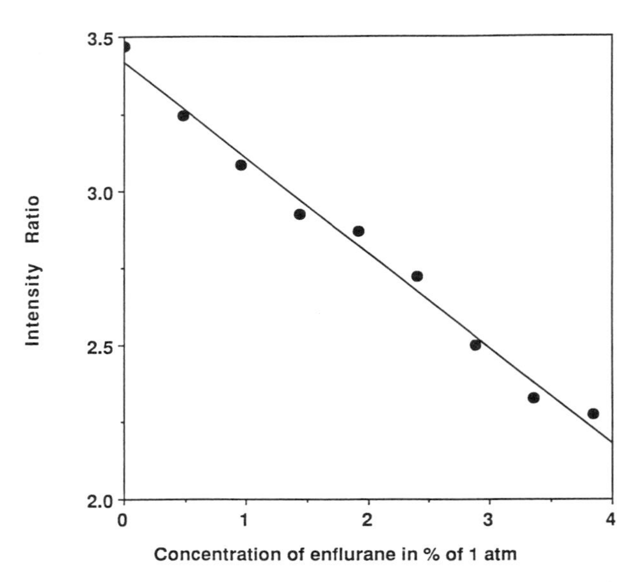

Fig. 5. Response of a probe made of DPPC/DMPC lipid mixture (90:10) at 38.5°C to increasing partial pressures of the anesthetic enflurane in the gas phase. 1 MAC (the alveolar concentration at which 50% of patients respond to surgical stimuli) for enflurane is 1.68%. The straight line is a least squares fit to the data.

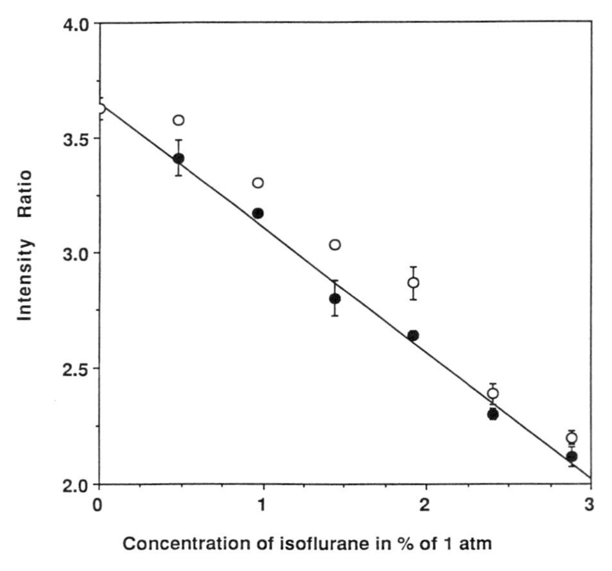

Fig. 6. Response of a probe made of DPPC/DMPC lipid mixture (86:14) at 38.2°C to increasing partial pressures of the anesthetic isoflurane in the gas phase: ○ after 15 min.; ● after 30 min.. 1 MAC for isoflurane is 1.15%. The straight line is a least squares fit to the 30 min data. $y = 3.64 - 0.54x$, $R^2 = 0.994$.

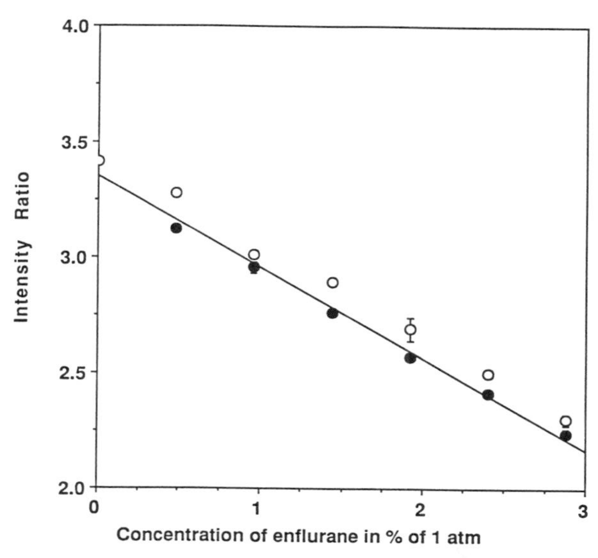

Fig. 7. Response of a probe made of DPPC/DMPC lipid mixture (86:14) at 38.1°C to increasing partial pressures of the anesthetic enflurane in the gas phase: ○ after 15 min.; ● after 30 min.. The straight line is a least squares fit to the 30 min data. $y = 3.35 - 0.40x$, $R^2 = 0.993$.

relevant concentrations of isoflurane at the physiological temperature of 37°C. The results of this experiment are presented in Fig. 8; again, linearity of R *versus* anesthetic partial pressure was seen over the range of pressures tested.

The effect of halothane was tested with a pure DPPC probe by adding four 15 μl aliquots of halothane and measuring R after each injection. Unfortunately, halothane, which, because of its bromine substituent, is well-known to be an efficient quencher of fluorescence, had a dramatic quenching effect on the Laurdan fluorescence. The lifetime of the probe, as determined by how long a usable SNR was obtainable, was, therefore, quite short. It is not yet certain whether the recorded changes in R are due to a fluidity change in the lipid or to differential quenching of Laurdan in the gel and liquid crystal bilayer forms. Further work is needed to solve this question.

Response to ethanol

Ethanol can also produce general anesthesia, and is known to depress the phase transition temperature of phospholipids [12–14]. The range of 0.0–5.0 mg of ethanol per ml of blood is of clinical interest. Higher concentrations produce coma and death. A probe made of DPPC/DMPC 86:14 molar ratio was immersed in water in the small vial described and, after thermal equilibration, exposed to increasing concentrations of ethanol in water. The vial, which contained a small stirring bar and the probe, was warmed in the thermostat bath (the temperature of the water in the bath was 38.3°C) until the temperature (in the vial) was around 38.1°C and the intensity ratio was approximately 3.55. The vial was then briefly removed from the bath, ethanol was added, the solution was stirred for one min, and the vial was placed back into the bath. Final concentrations of 2.5 and 5.0 mg of ethanol per ml of water were tested. The results of this experiment are reported in Fig. 9: small decreases in the ratio were observed for both concentrations. The small size of the change was expected given that the lipid phase transition is less sensitive to ethanol than to agents such as isoflurane and enflurane. It is interesting to note that there was no difference between the measurements after 15 and after 30 min, perhaps because of faster diffusion of ethanol through the gel, and the active stirring of the solvent.

CONCLUSIONS

The described sensor can potentially detect clinically relevant concentrations of general anesthetics in gas, blood, and tissue. Due to the intrinsic properties of fiber optic probes, the sensor features remote sensing capability, flexibility, small size, rapid response, and absence of electrical wires in the sensing region. Thus, this sensor has great potential for *in vivo* continuous monitoring of anesthetic concentrations. Direct extension of the work in this paper to a clinically useful device would necessitate the use of 4 optical fibers and two adjacent probes a *sample* probe in intimate contact with the environment, and a *reference* probe in thermal contact but chemically shielded to measure the local temperature. The anesthetic concentration would then be determined by comparison of the values of R from the two probes. Reduction to a single fiber for each probe is easily accomplished, as are a number of optical and/or electronic improvements of the current design. For example, as the measurements are based on the ratio of intensities at two wavelengths, the detection of fluorescence can easily be performed by two photodiodes and the optical analysis system can be a dichroic beam splitter. It may even ultimately be possible to reduce the system to one probe attached to one fiber.

Many problems remain to be solved before a device will be available in the clinic. Challenges currently common to all types of sensors, and particularly for biosensors, are reduction of response time, determination of biocompatibility, demonstration of reproducibility, and avoidance of degradation. To be determined are the effects of

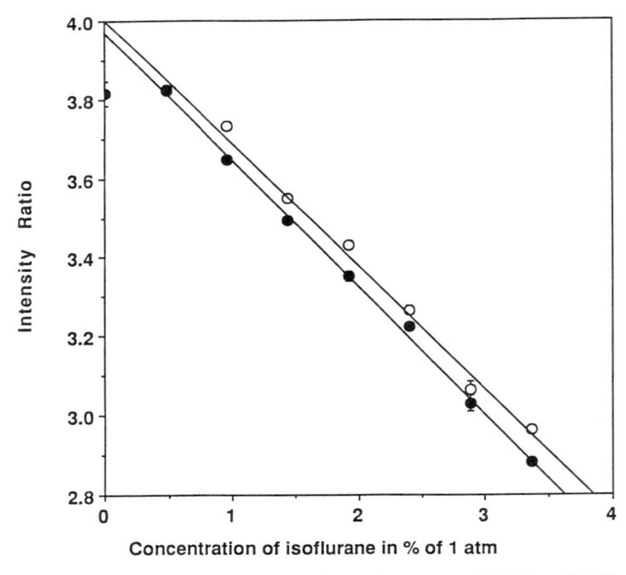

Fig. 8. Response of a probe made of DPPC/DMPC lipid mixture (80:20) at 37°C to increasing partial pressures of the anesthetic isoflurane in the gas phase: ○ after 15 min., $y = 4.00 - 0.32x$, $R^2 = 0.993$; ● after 30 min. $y = 3.97 - 0.32x$, $R^2 = 0.998$.

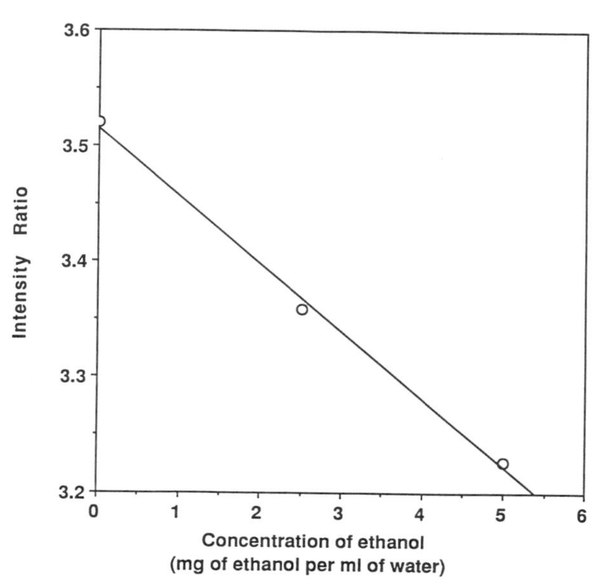

Fig. 9. Response of a probe made of DPPC/DMPC lipid mixture (86:14) immersed in water at 38.1°C to ethanol in the range 0–5 mg of ethanol per ml water. The straight line is a least squares fit to the data. $y = 3.51 - (5.86 * 10^{-2})x$, $R^2 = 0.997$.

sensor fouling on the host as well as of the effects of fouling on the sensor response time. Fortunately, an anesthetic sensor need be used *in vivo* for no longer than a day, during which time some of the deleterious effects of biofouling may not be a problem.

This study established two critical weaknesses in the sensor as currently designed: reproducibility of the results and degradation of the sensing element. The reproducibility of R values at a given temperature from probe to probe has proven to be dependent on the preparation of liposome populations with reproducible size distribution, as was discussed in our previous work [5]. This technological problem has been largely solved in the liposomal drug delivery industry, where the requirement for batch production of uniformly sized injectable particles is critical. Another worrisome factor is the possibility of hysteresis in the thermal behavior of the lipids. We are not yet certain how serious this problem will be, but have discussed (above) possible solutions if it proves to be unavoidable .

The primary reason for degradation of the sensing element *in vitro* is photobleaching of the fluorescent dye due to prolonged exposure to ultraviolet light. Several techniques are available for reduction of this potential problem. Note that greater light intensities were used when the probes were first illuminated than were necessary to produce acceptable SNR in these studies. Some of the problems associated with probe degradation are eliminated by ratiometric detection of fluorescence intensity. The response of this sensor does not depend on the absolute value of the intensity, as long as a usable SNR is obtained. Degradation of the sensing element may also occur by leaching of the dye out from the lipid and alterations of the lipid system. An appropriate barrier can be constructed that will allow sensing but avoid these problems.

Several attempts have recently been made to use lipid bilayers in optically based chemical sensors [15–18]. This is the first work of that we are aware of which the shift of a phase transition caused by introduction of an analyte is used as the operating principle of a fiber optic sensor. The principle of phase transition sensing is applicable to a variety of systems; the phospholipids used in this sensor are only an example of the many possible materials with phase transitions that could be used to sense the presence of analytes. We are pursuing extension of this approach to other detection schemes, transducing materials, and analytes.

ACKNOWLEDGMENTS

Much of this work was performed in the Optical Waveguide Chemical Sensors Laboratory of the Center for Process Analytical Chemistry of the University of Washington. Support has been provided by the Center for Bioengineering, the Center for Process Analytical Chemistry, the Rotary Foundation, and the Washington Technology Center. The work described in this manuscript is the subject of a patent filed with the U.S. Patent Office #07/367,508. S. Merlo's current address is: Universita' di Pavia, Facolta' di Ingegneria, Dipartimento di Elettronica, Laboratorio di Elettroottica, via Abbiategrasso 209, 27100 Pavia, Italy.

REFERENCES

1. R. D. Miller (Ed.) *Anesthesia*, Churchill Livingstone, 1981.
2. O. S. Wolfbeis and H. E. Posch. Fiber optical fluorosensor for determination of halothane and/or oxygen *Anal. Chem.* **57**:2556–2561 (1985).
3. M. G. Drexhage and C. T. Moynihan. Infrared optical fibers. *Sci. Am.* **11**:110–116 (1988).
4. H. E. Posch, O. S. Wolfbeis, and J. Pusterhofer. Optical and fibre-optic sensors for vapours of polar solvents. *Talanta* **35**(2):89–94 (1988).
5. S. Merlo and P. Yager. Optical method for monitoring the concentration of general anesthetics and other small organic molecules–An example of phase transition sensing. *Anal. Chem.* **62**:2728–2735 (1990).
6. S. Merlo, L. W. Burgess, and P. Yager. *Sensor Actuator* **A21-A23**, 1150–1154 (1990).
7. P. N. Yi and R. C. MacDonald. Temperature dependence of optical properties of aqueous dispersions of phosphatidylcholine. *Chem. Phys. Lipids* **11**:114–134 (1973).
8. T. J. J. Blanck. A simple closed system for performing biochemical experiments at clinical concentrations of volatile anesthetics. *Anesth. Analg.* **60**(6):435–436 (1981).
9. P. Yager and W. L. Peticolas. The kinetics of the main phase transitions of aqueous dispersions of phospholipids induced by pressure jump and monitored by Raman spectroscopy. *Biochim. Biophys. Acta* **688**:775–785 (1982).
10. S. Abrams and P. Yager. Manuscript in preparation.
11. A. S. Janoff and K. W. Miller. A critical assessment of the lipid theories of general anaesthetic action. In D. Chapman (ed.), *Biol. Membr.*, Academic Press, London, 1982, pp. 417–477.
12. E. S. Rowe. The effects of ethanol on the thermotropic properties of dipalmitoylphosphatidylcholine. *Mol. Pharmacol.* **22**:133–139 (1982).
13. E. S. Rowe. Lipid chain length and temperature dependence of ethanol-phosphatidylcholine interactions. *Biochemistry* **22**(14):3299–3305 (1983).
14. E. S. Rowe. Thermodynamic reversibility of phase transitions. Specific effects of alcohols on phosphatidylcholines. *Biochim. Biophys. Acta* **813**:321–330 (1985).
15. U. J. Krull, C. Bloore, and G. Gumbs. Supported chemoreceptive lipid membrane transduction by fluorescence modulation: the basis of an intrinsic fibre-optic biosensor. *Analyst* **111**(2):259–261 (1986).

16. U. J. Krull, R. S. Brown, R. F. DeBono, and B. D. Hougham. Towards a fluorescent chemoreceptive lipid membrane-based optode. *Talanta* **35**(2):129–137 (1988).

17. B. P. H. Schaffar and O. S. Wolfbeis. New optical chemical sensors based on the Langmuir-Blodgett technique. *SPIE Proc.*, 1988, pp. 990–18.

18. B. P. H. Schaffar, O. S. Wolfbeis, and A. Leitner. Optical sensors. Part 23. Effect of Langmuir-Blodgett layer composition on the response of ion-selective optrodes for potassium, based on the fluorimetric measurement of membrane potential. *Analyst* **113**:693–697 (1988).

A BAYESIAN KINETIC CONTROL STRATEGY FOR CYCLOSPORIN IN RENAL TRANSPLANTATION

Brian Whiting, Alison A. Niven, Andrew W. Kelman, Alison H. Thomson

Department of Medicine and Therapeutics
University of Glasgow

Janet Anderson, Angela Munday

Department of Pharmacy
Western Infirmary, Glasgow

J. Douglas Briggs

Department of Renal Medicine
Western Infirmary, Glasgow

INTRODUCTION

Although cyclosporin has brought about a dramatic improvement in graft survival after transplantation, its use in this therapeutic setting may be associated with considerable control problems. Many clinically oriented studies have shown remarkable fluctuations in cyclosporin blood concentrations and this has presented a challenge primarily to those interested in pharmacokinetics. This presupposes, of course, that the fluctuations are in large part due to pharmacokinetic variability of one sort or another. There may well be other reasons. An excellent overview of the topic – including suggestions for one pharmacokinetic strategy – has been provided recently by Kahan and Grevel [1]. In response to the often quoted and observed wide range of cyclosporin concentrations, these authors stress that a dosing strategy that achieves *uniform* drug levels by compensating for pharmacokinetic variability is essential for the promotion of rational cyclosporin regimens. While their comments were directed at renal transplantation, there is no doubt that such comments are equally applicable to the use of cyclosporin in other transplant situations.

Control is necessary because the boundaries between graft survival, graft rejection and adverse reactions in terms of cyclosporin concentrations may be almost indistinguishable, yet the clinical response to concentration information is often quite ad hoc. This may reflect ambiguity about cyclosporin target concentration ranges – a reflection of an unfortunate paucity of data on cyclosporin concentration-response relationships. The picture is also confused – particularly in the immediate post-operative period when the drug is administered orally – by an almost unprecedented amount of day-to-day variability in cyclosporin concentrations.

We have studied this problem (in renal transplantation) over a number of years and have attempted to develop a control strategy based on Bayesian parameter estimation, using only limited amounts of clinical and laboratory information. Earlier parts of this work have been reported elsewhere [2–4].

Advanced Methods of Pharmacokinetic and Pharmacodynamic Systems Analysis
Edited by D'Argenio, Plenum Press, New York, 1991

INITIAL PHARMACOKINETIC STUDY

The pharmacokinetics of cyclosporin following renal transplantation were studied intensively in 11 patients for periods of up to 60 post-operative days. Because of patient limitation vis-à-vis blood sampling over short time periods, a Bayesian approach [5] was used to estimate clearance (Cl) and volume of distribution (V) values in each patient throughout the study period. This yielded a profile of the pharmacokinetic behavior of each patient which extended from the immediate post-operative period to approximately 8 weeks. The Bayesian procedure employed a traditional time weighting function (progressively down-weighting the importance of past concentration data) and assumed a (relatively) constant clearance (and volume of distribution).

On average, there were 33 sets of Bayesian parameter estimates per patient over the 60 day period. When these were viewed consecutively as a function of time, there was a striking consistency about the way they behaved. For example, a plot of Cl versus time (recognizing that Cl is, in fact, the ratio of clearance to bioavailability, or Cl/F) revealed that Cl decreased progressively with time and asymptotically approached a relatively constant baseline value. The shape of the decline (Fig. 1) suggested that a simple exponential function would explain it best, and this idea was pursued with the NONMEM software [6] because of the repeated nature of the observations and the underlying nonlinear function. Moreover, NONMEM would allow quantitation of the variability associated with the parameters (a key issue in this work).

The analysis revealed the following relationship:

$$Cl/F = 36.2e^{-0.147t} + 14.3 \tag{1}$$

where Cl/F declined (exponentially) from an average initial value of 50.5 l/h at a rate of 0.147 day^{-1} to a final, average ("constant") value of 14.3 l/h.

In relation to the usual kind of analysis carried out with NONMEM, the data set here was unusually rich in terms of the amount of information per subject but poor in the number of subjects. Estimates of intersubject variability, therefore, may well

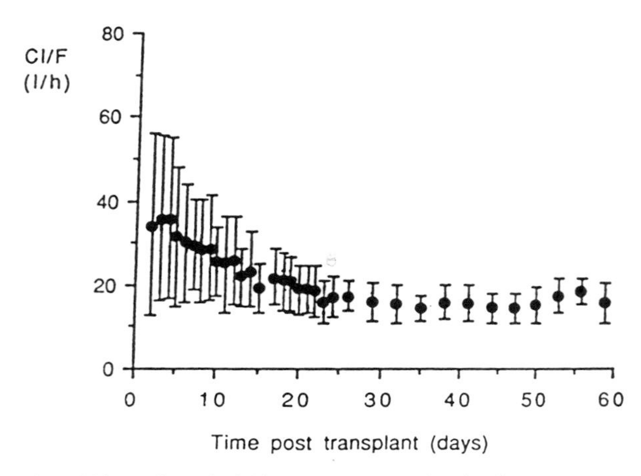

Fig. 1. Mean successive Cl/F values (±SD) in 11 patients in the first two months following renal transplantation.

reflect this deficiency in numbers, but do provide a starting point for subsequent Bayesian parameter estimation (see below). This essentially statistical drawback, combined with the evident variability in blood levels, gave estimates of variability (coefficients of variation) in initial Cl/F, rate of decline and asymptotic Cl/F of 50%, 103% and 20%, respectively.

FIRST BAYESIAN EVALUATION

Traditional Bayesian parameter estimation assumes that there will be no consistent changes in the pharmacokinetics from point (one concentration) to point (the next concentration). Kinetics will, however, change with alterations in clinical status and such alterations are allowed for by downgrading the influence of previous concentration measurements so that they have a diminishing influence on the maximum likelihood procedure employed. This allows the Bayesian estimates to shift to accommodate new information. This device has been referred to as a "forgetting memory". In the present circumstances, however, the Bayesian procedure was modified to take account of the fact that consecutive concentrations *should* provide information about the change in pharmacokinetics and indeed the above exponential function was embedded into the underlying structural pharmacokinetic model so that estimates of initial, and final, clearances could be obtained within individual patients as well as the associated (exponential) rate of decline.

A further 18 patients were then used to evaluate this modification, in competition with the original (constant clearance) procedure. Moreover, two contending underlying pharmacokinetic models were tested – the one and two compartment models. Thus, the second study tested four Bayesian strategies:

1. One compartment model: constant Cl/F
2. One compartment model: CHANGING Cl/F
3. Two compartment model: constant Cl/F
4. Two compartment model: CHANGING Cl/F

To gain an overview of any differences in the predictive performance of these four procedures, concentration-time data from the first two post-operative weeks were used to "predict" concentrations in the second or third two week periods (i.e., weeks 3 and 4 (time A) and 5 and 6 (time B)) as shown in Fig. 2. The predictions were made retrospectively on the basis of carefully recorded dosage histories and the prediction times were chosen to coincide with the times of actual cyclosporin concentration measurements. Thus, at each time, a prediction error could be calculated from the difference between observed and predicted concentrations. The effect of distance (i.e., time) from the original data (weeks 1 and 2) and the complexity of the underlying model (and procedure) is shown in Table I [4]. Although these data are presented parametrically, analysis was carried out with a Friedman ANOVA and showed a significant reduction in the bias associated with procedures 2 and 4, i.e., procedures incorporating the exponentially changing Cl/F function. Precision throughout, however, remained relatively poor with an average standard deviation of about 60μg/l. This implies that even at best – using procedure 2 to make dosage adjustments for the third and fourth post-operative weeks – a target concentration of, say, 175μg/l could only be obtained within a 68% confidence interval of 115–235μg/l. This, however, must be seen in the context of uncertainty about the extent

Table I. Mean Prediction Errors (μg/l) and their Standard Deviations

Model/ Procedure	Time A	Time B
1	70±56	100±55
2	17±51*	26±53*
3	82±59	110±61*
4	26±65*	48±71*

* $p < 0.001$ Friedman ANOVA.

of the "therapeutic window" or therapeutic range applicable to individual patients. Various values for this have been quoted and the target itself may change with time, cyclosporin requirements apparently diminishing progressively in the months following operation. Therapeutic ranges quoted have a value, on average, of about 150–200 concentration units, centered on mean targets of the order of 200μg/l. This implies that the performance of the Bayesian system – as it behaved in this evaluation – would not be entirely satisfactory from a dosage adjustment point of view. The interval between observation and prediction, however, can be considerably less than 1–2 weeks, and the conditions set up in this evaluation were probably too severe when viewed pragmatically.

SECOND BAYESIAN EVALUATION

With the above caveat in mind, a further – prospective – evaluation study has been set up, where the prediction intervals are much shorter, ideally spanning the time between consecutive concentrations (say a minimum of 3 days). So far, data from 26 new renal transplant patients have been analyzed, where prediction errors have again been calculated using procedure 2. The prediction interval has, in fact, been shortened to no more than 7 days. Concentrations in the second post-operative week have been predicted from the first week's data and concentrations in week 3 have been predicted from all data in the first 2 weeks (an average of 3 concentrations

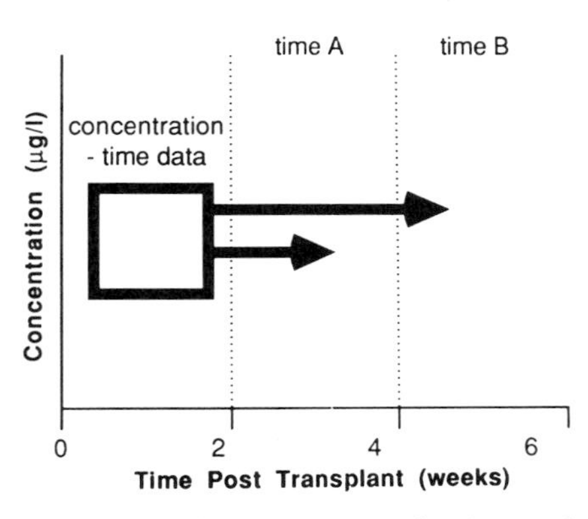

Fig. 2. Plan of first Bayesian evaluation study.

per week). The results, now expressed as percentage prediction errors (the error expressed as a percentage of the observed concentration) are shown in Fig. 3. While there are some extreme values, and the mean percentage prediction error is still biased positively, most are in a band extending from 0 to 100 and definite improvement can be seen (as might be expected) when concentrations in the third week were predicted from all available data in the first two weeks. The improvement is not so much in the bias (this remains) but in the spread (precision) of prediction errors.

DISCUSSION

It is apparent from the results presented here that while an attempt has been made to account for some of the variability in cyclosporin pharmacokinetics after renal transplantation, the extra sophistication provided by the "changing clearance function" still fails to make a real impact on the variability in the relatively early post-operative period. Others have reported on the consistency of the change in pharmacokinetics and have attributed it to a real decrease in clearance (relative to the apparently high initial levels), a real increase in bioavailability (a function of intestinal absorption in the post-transplant period) or to a combination of these factors. It seems sensible, therefore, to account for this shift in pharmacokinetics and to include estimates of the rate of change in Cl/F in the derivation of concentration predictions, thereby increasing the efficiency with which dosage adjustments can be made to achieve specific target concentrations. That this has not yet been more successful means that the time dependent change in pharmacokinetics is only one component of the variability which is so much in evidence. Other reasons for this variability must now be searched for and great attention must be paid to the detail of patient care, the recording of day-to-day post-operative condition (particularly gastrointestinal disturbances) and drug administration. We still feel that a Bayesian approach is valuable in these circumstances, but it must be coupled with careful

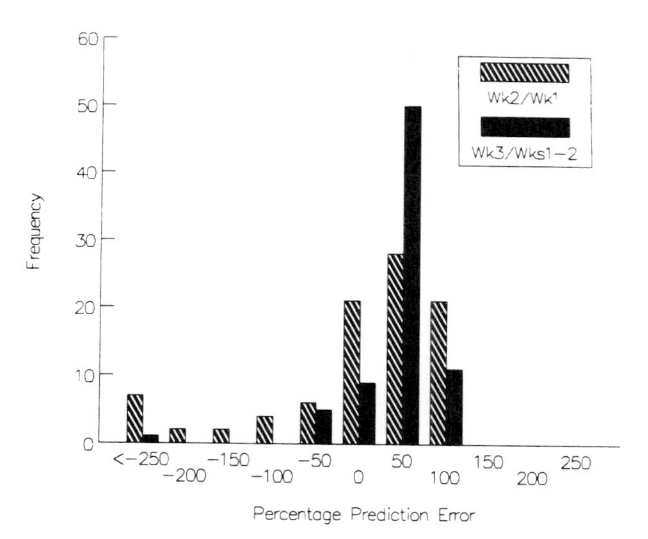

Fig. 3. Second evaluation study: percentage prediction errors in 26 patients. Week 2 concentrations predicted from week 1 (shaded bars); Week 3 concentrations predicted from weeks 1 and 2 (solid bars).

clinical observation and meticulous attention to drug administration procedures and the timing of blood samples.

REFERENCES

1. B. D. Kahan and J. Grevel. Optimization of cyclosporin therapy in renal transplantation by a pharmacokinetic strategy. *Transplantation* **46**:631–644 (1988).
2. A. A. Niven, J. Grevel, M. Al-Banna, A. W. Kelman, B. Whiting, and J. D. Briggs. Pharmacokinetics of cyclosporin in the early post-operative period following renal transplantation. *Brit. J. Clin. Pharmaco.* **26**:626 (1988).
3. B. Whiting, A. A. Niven, A. W. Kelman, and J. D. Briggs. Improved therapeutic control of cyclosporin in renal transplantation. *Decision Support for Patient Management: Measurement, Modeling and Control,* British Medical Informatics Society, London, 1989, pp. 33–38.
4. A. A. Niven. The control of cyclosporin in transplantation: Pharmacokinetic aspects. Doctoral Thesis, University of Glasgow, 1990.
5. A. W. Kelman, B. Whiting, and S. M. Bryon. OPT: A package of computer programs for parameter estimation in clinical pharmacokinetics. *Brit. J. Clin. Pharmaco.* **14**:247–256 (1982).
6. S. L. Beal and L. B. Sheiner. *NONMEM User's Guides,* I–VI. University of California, San Francisco, 1979–1989.

TARGETED SYSTEMIC EXPOSURE FOR PEDIATRIC CANCER THERAPY

John H. Rodman and **William E. Evans**

Pharmacokinetics and Pharmacodynamics Section
Pharmaceutical Division
St. Jude Children's Research Hospital
and
The Center for Pediatric Pharmacokinetics and Therapeutics
Departments of Clinical Pharmacy and Pediatrics
University of Tennessee, Memphis

INTRODUCTION

Pharmacokinetic variability in pediatric patients due to maturational changes in organ function, effects of concomitant disease, and drug toxicity or interactions commonly results in drug clearances that differ by a factor of 4 or 5. This intersubject pharmacokinetic variability has been shown to correlate to an increased likelihood of toxicity in patients with low drug clearances, and therapeutic failure in patients with high drug clearances [1, 2]. Pharmacokinetic and pharmacodynamic modeling strategies have been developed and incorporated into clinical studies intended to define the unique pharmacokinetics of anticancer drugs in pediatric patients, identify clinical correlates (e.g., patient characteristics, laboratory indices of organ function) of pharmacokinetic differences, and adjust dosage regimens to control for pharmacokinetic variability.

The initial step in modeling drug disposition is to establish a population pharmacokinetic model for the drug. Estimation of population pharmacokinetic parameters is most often undertaken during drug development to provide the basis for dosage regimen design or subsequently to compare subgroups of patients for the influence of factors such as age, organ dysfunction (e.g., renal failure). Recently developed population modeling methods based on limited sampling strategies (3 to 5 samples per patient) offer substantial practical advantages over conventional methods based on more extensive sampling strategies. An efficient population estimation strategy is particularly pertinent for the pediatric population where data per subject is limited and to permit extending studies to larger and thus, more representative populations. A summary of a recent comparison we have completed provides some important insight into the performance of population parameter estimation methods.

The population model provides the basis for establishing initial dosage regimens and also the framework for estimation strategies that can be used to tailor dose regimens to specific target concentration profiles. Parameter estimation in the individual patient has been extensively used for drugs with a narrow therapeutic range such as cardiovascular drugs and anticonvulsants but has not been widely applied to anticancer therapy. In addition to a potentially important role in optimizing therapy for individual patients, the utility of controlling pharmacokinetic variability during Phase I-III studies has been examined [3]. A clinically feasible strategy using a Bayesian estimation algorithm has been developed for anticancer drugs such as methotrexate and teniposide and prospectively evaluated in a clinical study. The results demonstrate the potential utility of wider application of such strategies both during drug development and in routine clinical care.

TARGETED SYSTEMIC EXPOSURE FOR ANTICANCER DRUGS

When fixed doses of anticancer drugs are given to patients, a uniform response (e.g., survival, hematological suppression) is confounded by the underlying biological variability of the host and disease as well as pharmacokinetic and pharmacodynamic variability (Fig. 1). Pharmacokinetic differences among patients is a particularly relevant source of variability in that by adjusting dosage regimens, it can be controlled. In contrast, the underlying biology of the tumor or the tolerance of the patient to therapy are characteristics much less subject to intervention. It is worth noting, as an aside, that with the advent of recombinant biotechnology and the availability of cytokines, such as the growth factors, the biology of the host and tumor interface is increasingly a target for therapeutic intervention as well.

Pharmacokinetic modeling provides a framework for describing the variability in drug disposition between and within patients that results in substantial differences in systemic exposure. Systemic exposure and its variability could be quantified with a number of variables such as systemic clearance, maximum concentration, duration of time above a putatively effective concentration, or a time averaged concentration.

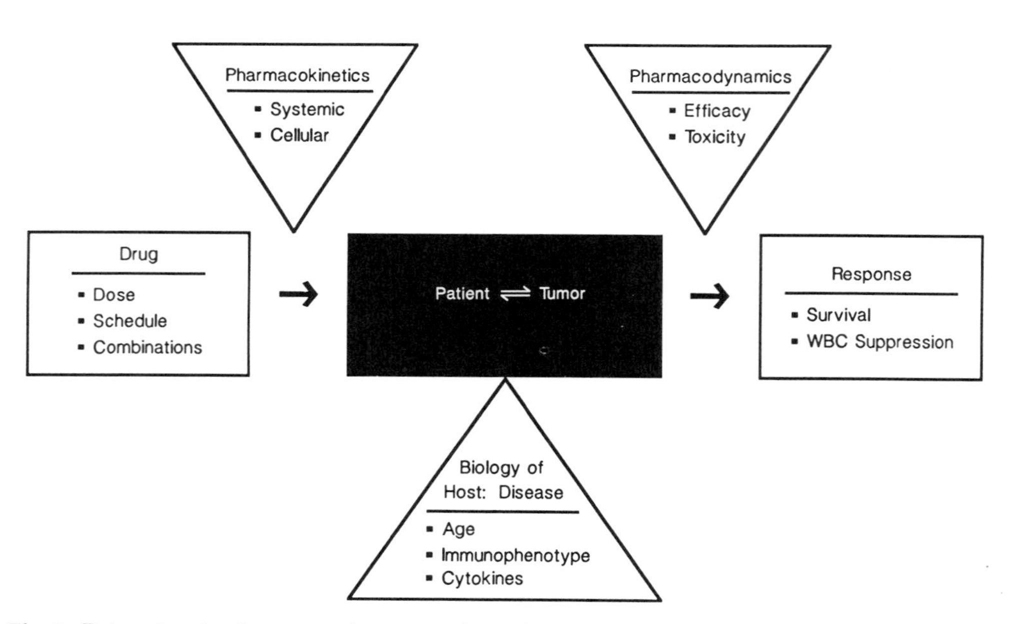

Fig. 1. Determinants of response for cancer chemotherapy. Variables such as dose, schedule, and therapeutic combinations can be manipulated in an effort to optimize response (e.g., survival, hematological toxicity). However, pharmacokinetic and pharmacodynamic variability among patients are treated with the same regimen.

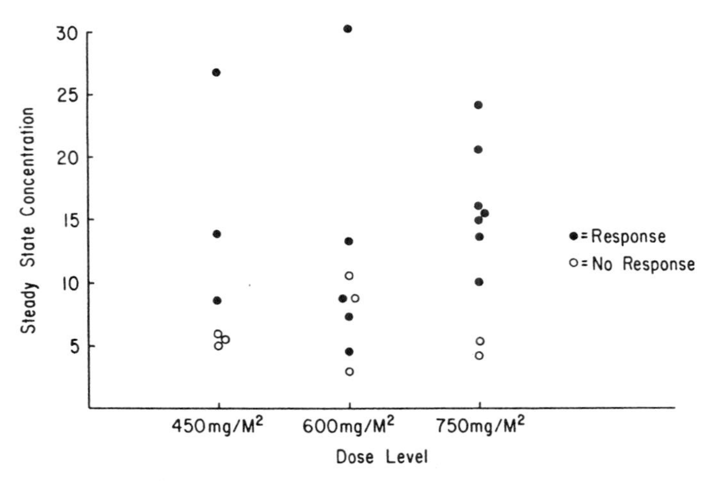

Fig. 2. Steady-state VM26 concentrations and responses. Doses of teniposide (VVM26) adjusted for body size yielded steady state drug concentrations following 72 hours continuous infusions that varied by a factor of 5 or 6.

The area under the plasma concentration time curve (AUC) is an intuitively appealing measure of systemic exposure with supporting data from the laboratory [4] and clinically relevant mathematical models [5]. Data from our laboratory for several drugs [1,2] indicates that such relationships can be developed during Phase I-III studies in pediatric cancer.

The degree of interpatient pharmacokinetic variability is illustrated for the investigational drug VM-26, or teniposide in Fig. 2. Although the average concentrations rise proportionately with increases in dose, the interpatient variability obscures a dose effect relationship as patients receiving doses differing by almost a factor of 2 can have similarly high or low concentrations. Furthermore, when oncolytic response was measured in these patients, 5 of 13 subjects with steady state concentrations less than 12 mg/l failed to respond while 10 of 10 with values above 12 mg/l demonstrated an oncolytic response. When both efficacy and toxicity were examined using AUC as a measure of systemic exposure, the response profiles shown in Fig. 3 could be defined. Note that within a 3 fold range of systemic exposure, measured here as the AUC, patients with low values are unlikely to respond while patients with high AUC's are at substantial risk of toxicity.

THE POPULATION ESTIMATION PROBLEM

The most widely-used approach for summarizing the pharmacokinetic parameters for a population has been referred to as the Two Stage method [6]. Briefly, it entails conducting a series of studies in individual subjects with a significant number of blood (and much less often urine) samples obtained. For each subject an estimate of the pharmacokinetic parameters is determined, for example, using a compartmental model fit to the data with weighted nonlinear least squares [7]. These parameters are then summarized with conventional statistical methods and possible relationships between factors such as weight, age, and organ function are examined.

An alternative approach proposed by Beal and Sheiner [8, 9] directly estimates the population statistics (e.g., mean and variance) for the pharmacokinetic parameters such as body size and age within a nonlinear mixed effects model. Among the attractive attributes of this approach are the ability to use limited observations (i.e., 1 to 5) per subject but from large populations (i.e., several hundred) and the explicit incorporation of an error model to accommodate the random residual variability arising from factors such as model misspecification. The explicit estimation of the random residual error has the potential to reduce the upward bias in the estimate of interindividual pharmacokinetic variability inherent in the more widely used two stage method.

The application of the nonlinear mixed effects population models (one stage approach) for estimation of pharmacokinetic parameters has been proposed primarily for observational data collected during routine clinical therapy. The use of this approach for conventional intensive pharmacokinetic studies where there is more extensive data but in fewer subjects, has been suggested [8] but not extensively evaluated. It offers the potential benefit of providing more robust and reliable estimates of intersubject pharmacokinetic variability and the potential option of developing more efficient (i.e., fewer samples) study designs. Two possible shortcomings are the limitations of small sample size and the potential, carefully acknowledged [9] bias arising from the utilization of a first order approximation using a Taylor Series expansion to accommodate the nonlinear random interindividual effects in the population model.

To examine the reliability of the First Order method incorporated in a nonlinear mixed effects model (FO), a Monte Carlo simulation study was done based on an experimental design for the investigational anticancer drug teniposide. Previously obtained pharmacokinetic parameters for a two compartment model and estimates for intersubject variability were used to generate data sets of 12 measured drug concentrations for a population of 15 simulated subjects. Normally distributed, random "observation" error with a coefficient of variation of 10% was added to the nominal concentrations. This was repeated 10 times, thus, generating data sets corresponding to 10 separate studies of 15 subjects. Each of the studies was then analyzed using the FO method as implemented in the NONMEM computer package[9].

The results of the Monte Carlo study are summarized in Fig. 4 for the population estimate for teniposide clearance. The FO method was consistently and significantly biased. On average across the 10 studies, the mean estimate for clearance was only 72% of the true sample mean. A similar order of magnitude of bias was apparent for

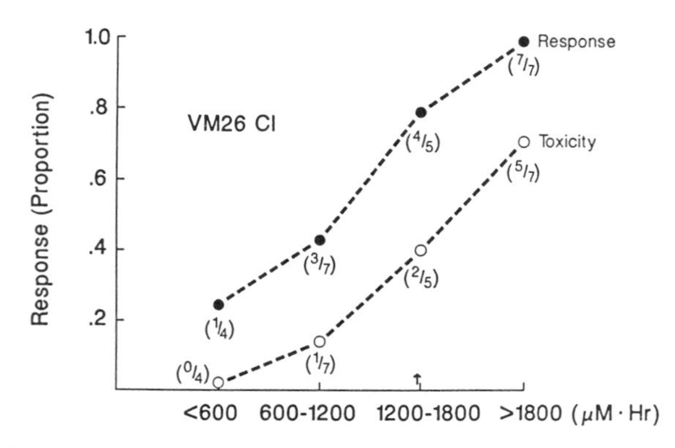

Fig. 3. Relationship between systemic exposure for teniposide (horizontal axis) and both toxicity and efficacy shown as proportion of patients (vertical axis).

72% of the true sample mean. A similar order of magnitude of bias was apparent for all structural model parameters. For purposes of comparison, the same data were analyzed with a standard two stage approach and there was significantly less bias. The mean clearance from the two stage estimates was 93% of true values. The FO method did provide reliable estimates of both intersubject variability and the random residual variability.

The magnitude of the bias for the FO method was somewhat surprising and disappointing. To clarify the possible impact of small sample size, the data from all 10 simulation studies were pooled and analyzed with the FO method and the bias remained similar for all estimates of mean structural model parameters as that found with small sample size. Differences between the results here and previously reported experience with the FO method [8] may arise from the study design and the model structure. It may be that for higher order models with a larger number of nonlinear parameters under multiple dose, non-steady state conditions, the performance of the first order approximation is less satisfactory. Further refinement of this important estimation strategy appears needed before it can be reliably used for model structures and study designs such as evaluated here.

PARAMETER ESTIMATION IN THE INDIVIDUAL

Substantial pharmacokinetic variability can confound the evaluation of the consequences of combining drugs especially when sample size is small. The use of combinations of anticancer agents with complementary or synergistic mechanisms

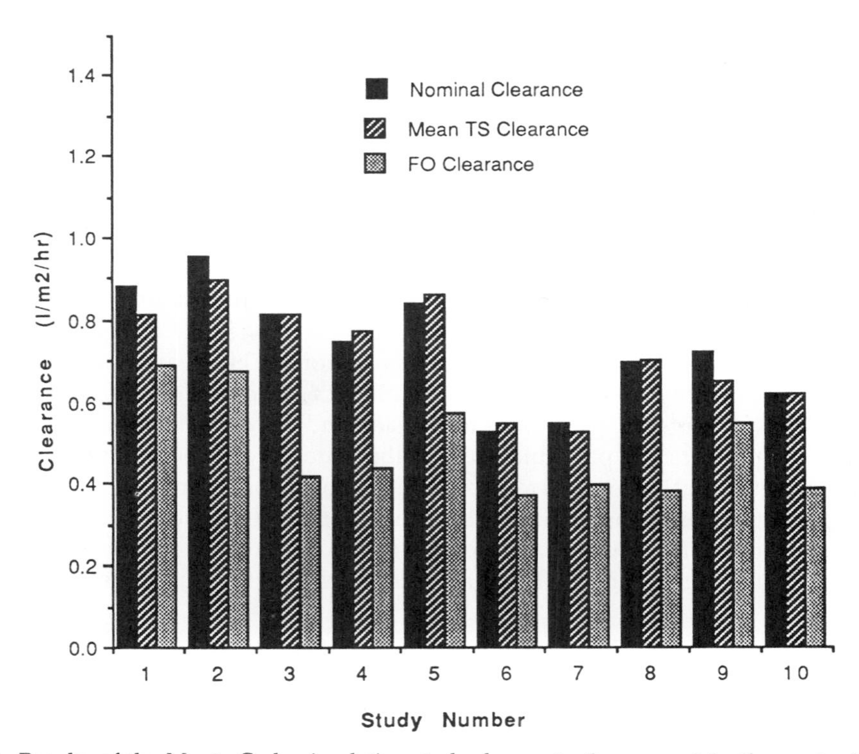

Fig. 4. Results of the Monte Carlo simulation study demonstrating a consistently greater bias for clearance of teniposide estimated with the First Order method.

ical benefit [5]. Studies have suggested enhanced intracellular accumulation of methotrexate and methotrexate polyglutamates in the presence of teniposide at concentrations that are achievable in patients. This cellular interaction and the well established independent antileukemic activity of both methotrexate and teniposide provided a compelling rationale for evaluating this previously untested combination of drugs and to prospectively evaluate a Bayesian estimation algorithm in pediatric cancer patients [3].

A continuous infusion regimen at doses of each drug known to be independently effective and well tolerated was selected to maximize the potential cellular interaction between teniposide and methotrexate. To control for interpatient pharmacokinetic variability as a confounding factor in the evaluation of these agents, we developed a pharmacokinetic strategy for design of patient specific dosage regimens that would permit us to achieve target plasma concentrations. The specific study objectives were to determine the pharmacokinetics of methotrexate and teniposide given alone or simultaneously and to evaluate the precision and demonstrate the feasibility of a dose optimization strategy for these two important anticancer agents.

Pharmacokinetic studies of methotrexate and teniposide were conducted in 19 children with relapsed acute lymphocytic leukemia. Patients were randomly assigned to receive methotrexate as a 24 hr continuous infusion either simultaneously with a continuous infusion of teniposide or sequentially with the teniposide infusion beginning 12 hours after the end of the methotrexate infusion. Plasma samples for measurement of methotrexate (n=12) and teniposide (n=11) were obtained during and after infusions at appropriate times for a comprehensive pharmacokinetic study of each drug. Two measured drug concentrations obtained at 1 and 6 hours after the start of the infusion were used to adjust the dose rate for each patient to achieve target values of 10μM for methotrexate and 15 μM for teniposide. The dose adjustments were made before hour 12 of therapy.

Pharmacokinetic parameters for teniposide were not different in patients receiving the simultaneous and sequential therapy and the median value for clearance (13.4 ml/min/m) was similar to previous studies. Despite similar end of infusion methotrexate concentrations (simultaneous group–9.66 μM; sequential group–9.72 μM) the 24 hour post-infusion MTX concentrations were lower (0.137 vs 0.230 μM; $p<.05$) in the patients receiving simultaneous infusions consistent with teniposide enhancement of intracellular methotrexate. Despite substantial interpatient pharmacokinetic variability, the patient specific dose regimens yielded acceptably precise and minimally biased steady state drug concentrations. The mean model-predicted end of infusion concentrations were 9.69μM (97% of target, CV=17%) for MTX and 16.13μM (108% of target, CV=35%) for teniposide and were similar in sequential and simultaneous patients. Of particular note in this study is that parameter estimation was accomplished using only two observations for a four parameter model with sampling times prior to the attainment of steady state. This indicates that with an appropriately constructed Bayesian prior, the revised (posterior) parameter estimates are sufficiently precise for control despite a clearly suboptimal design strategy.

SUMMARY

Pharmacokinetic variability among pediatric patients receiving anticancer drugs is substantial. The use of innovative population estimation methods and strategies for estimation and control in the individual subject can play a major role in determining factors associated with variability in drug disposition and minimizing this mining factors associated with variability in drug disposition and minimizing this as a contributing factor to therapeutic failure or toxicity. Further work is needed to provide more reliable and informative population estimation methods and control strategies that more explicitly incorporate pharmacodynamic endpoints.

ACKNOWLEDGMENT

These studies were carried out with the diligent support of the clinical pharmacy, medical and nursing staff at St. Jude Children's Research Hospital and the cooperation of the patients and parents whom we strive to help. Supported in part by grants from the National Institutes of Health: CA20180 (Leukemia Program Project Grant), R37 CA36401, CA217865 (Core Cancer Center Grant); a Center of Excellence grant from the State of Tennessee, and the Syrian-American Lebanese Associated Charities.

REFERENCES

1. W. E. Evans, W. R. Crom, M. Abromowitch, *et al.* Clinical pharmacodynamics of high dose methotrexate in acute lymphocytic leukemia. *N. Engl. J. Med.* **314**:471–477 (1986).
2. J. H. Rodman, M. Abromowitch, J. A. Sinkule *et al.* Clinical Pharmacodynamics of Continuous Infusion Teniposide: Systemic Exposure as a Determinant of Response in a Phase I trial. *J. Clin. Oncol.* **7**:1007–1014 (1987).
3. J. H. Rodman, M. Sunderland, R. L. Kavanagh, J. Ochs, J. Yalowich, W. E. Evans, and G. K. Rivera. Pharmacokinetics of continuous infusion methotrexate and teniposide in pediatric cancer patients. *Cancer Res.* **50**:4267–4271 (1990).
4. R. S. Day. Treatment sequencing, asymmetry, and uncertainty: protocol strategies for combination chemotherapy. *Cancer Res.* **46**:3876–3885 (1986).
5. A. J. Coldman, C. M. L. Coppin, and J. H. Goldie. Models for Dose Intensity. *Math. Biosci.* **92**:97–113 (1988).
6. C. C. Peck and J. H. Rodman. Analysis of clinical pharmacokinetic data for individualizing drug dosage regimen. In W. E. Evans, J. J. Schentag, and W. J. Jusko (eds.), *Applied Pharmacokinetics: Principles of Therapeutic Drug Monitoring*, Applied Therapeutics, Inc., Spokane, 1986, pp. 55–82.
7. D. Z. D'Argenio and A. Schumitzky. A program package for simulation and parameter estimation in pharmacokinetic systems. *Comp. Prog. Biomed.* **9**:115–134 (1979).
8. S. L. Beal. Population pharmacokinetic data and parameter estimation based on their first two statistical moments. *Drug Metab. Rev.* **15**:173–194 (1984).
9. S. L. Beal and L. B. Sheiner. *NONMEM User's Guides*, NONMEM Project, University of California, San Francisco, 1989.

TARGETING THE EFFECT SITE WITH A COMPUTER CONTROLLED INFUSION PUMP

Steven L. Shafer

Department of Anesthesia
Stanford University School of Medicine
and
Palo Alto Veterans Administration Medical Center

ABSTRACT

Over the last decade computer controlled infusions have been used to rapidly achieve and maintain constant plasma drug concentrations for drugs having pharmacokinetics characterized by polyexponential disposition functions. For most drugs the plasma is not the site of drug effect. Thus, targeting the plasma drug concentration may not be a rational approach to optimal drug therapy. Drawing on previously described *effect site* models, an algorithm is developed for computer controlled infusion pumps to target the apparent concentration at the site of drug effect, rather than the concentration in the plasma. Simulations are based on the pharmacokinetics and plasma-effect site equilibration delay that have been reported for fentanyl, an intravenous opioid commonly used in anesthetic practice.

INTRODUCTION

The pharmacokinetics of many intravenous drugs used in anesthesia are described by a three compartment mammillary model, as shown in Fig. 1. Drug is administered into and eliminated from the central compartment. The central compartment is also the site of blood sampling. Drug is also transferred from the central compartment into rapid and slow distributional compartments. Based on preliminary work by Krüger-Thiemer [1], Schwilden described a general method of rapidly reaching and maintaining a constant plasma drugs concentration for drug described by multicompartment mammillary models [2]. Many research groups have incorporated this linear model into computer controlled infusion pumps (CCIP) using a variety of algorithms to maintain steady plasma drug concentrations [3,4,5,6,7].

Figure 2 shows a simulated anesthetic course using a computer controlled infusion of fentanyl, a synthetic opioid, based on the pharmacokinetic parameters reported by Scott and Stanski [8]. Following induction of anesthesia with an intravenous hypnotic (e.g., thiopental), the desired fentanyl concentration to be maintained by the CCIP is set by the anesthesiologist at 6 ng/ml for endotracheal intubation, which is the most stimulating part of the anesthetic. Following intubation, the target concentration is decreased to 3 ng/ml, and then to 2.5 ng/ml. The target concentration is increased to 4 ng/ml immediately prior to incision in anticipation of

Advanced Methods of Pharmacokinetic and Pharmacodynamic Systems Analysis
Edited by D'Argenio, Plenum Press, New York, 1991

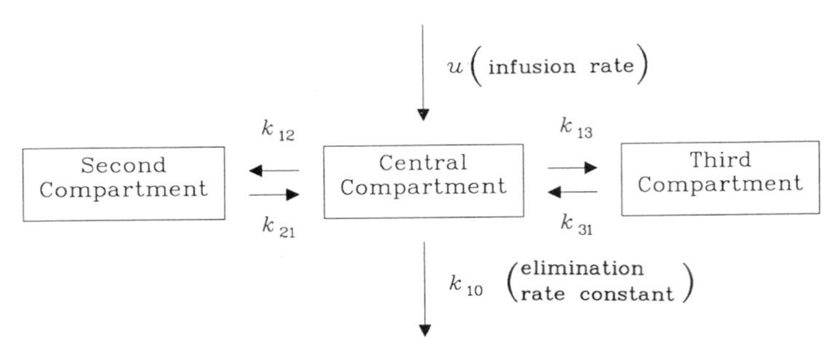

Fig. 1. Three compartment (i.e., triexponential) mammillary pharmacokinetic model.

the increased stimulation. Over the course of the surgery the target plasma fentanyl concentration is titrated downwards, based on patient responsiveness, to a concentration of 2 ng/ml. With the increased stimulation of skin closure, the fentanyl concentration is briefly increased to 2.5 ng/ml. Following skin closure, the infusion is terminated and the patient awakens from anesthesia.

Titrating an infusion to the plasma concentration, as shown in Fig. 2, is irrational for most of the drugs used in anesthesia because *the plasma is not the site of drug effect*. Sheiner *et al.* [9] and Hull *et al.* [10] independently proposed that the three compartment model be modified for drugs whose site of action is not the plasma by the addition of an effect compartment, as shown in Fig. 3. The effect compartment is postulated to have negligible volume, so it does not influence the pharmacokinetic model. The effect compartment model allows parametric characterization of the hysteresis between drug administration and drug effect. This hysteresis has been measured for many intravenous anesthetics. For fentanyl, the half-time of equilibration between the plasma and the site of drug effect is about 6 minutes [11]. Since the concentration at the site of drug effect cannot be measured directly, the concentration at the site of drug effect is expressed as the *apparent* concentration, which is the plasma concentration at steady state which would produce the same degree of drug effect.

Figure 4 shows the apparent fentanyl concentration at the site of drug effect produced by the dosing regimen from Fig. 2. The precise increases and decreases in

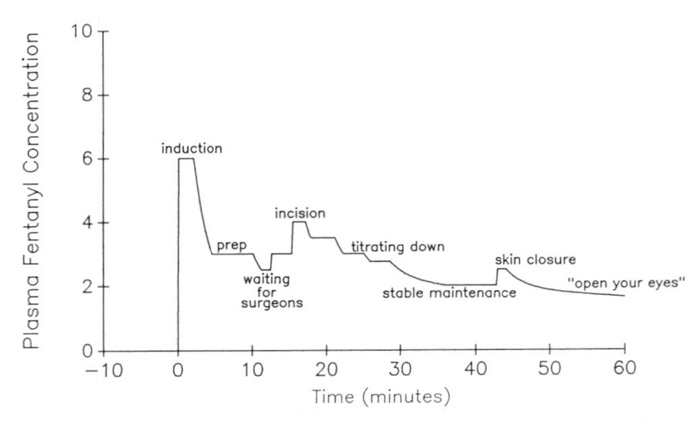

Fig. 2. Simulated plasma fentanyl concentration over time produced by a CCIP for a brief surgical procedure.

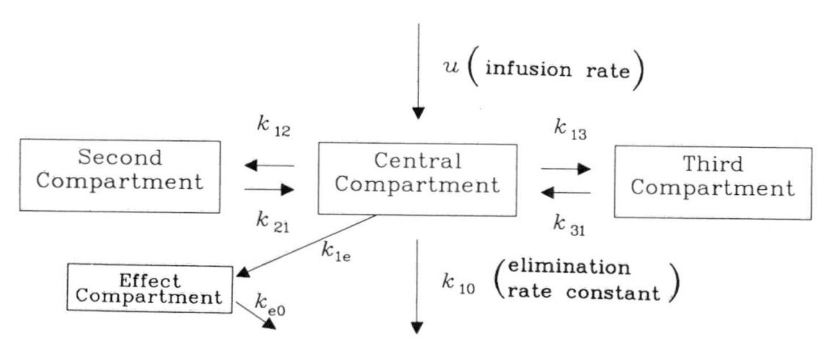

Fig. 3. Three compartment mammillary model with an additional effect compartment. The disposition of the effect compartment is proportional to $e^{-k_{e0}t}$.

plasma fentanyl concentration seen in Fig. 2 have produced a slurred rise and fall in apparent fentanyl concentration at the site of drug effect. The apparent concentrations at the site of drug effect only minimally reflect the desired target concentrations. This manuscript will describe an algorithm by which a CCIP can target the apparent drug concentration at the site of drug effect, rather than the concentration in the plasma, and thus, account for the equilibration delay between the plasma concentration and the apparent concentration at the site of drug effect.

ALGORITHM

Figure 5 shows the response of the model (as shown in Fig. 3) to a bolus injection. Following the bolus injection there is a very rapid decrease in plasma fentanyl concentration, primarily from redistribution of fentanyl into peripheral tissues. Concurrently, there is an increase in the apparent concentration at the site of drug effect, which peaks when the plasma concentration equals the apparent effect site concentration. Beyond this time point (approximately 3.75 minutes), both the plasma concentration and the apparent concentration at the site of drug effect decrease over time. The time of peak effect has been marked with a vertical line.

Provided there is no drug in the body, the time to peak concentration at the site of drug effect following a bolus injection is independent of the dose. This is a necessary

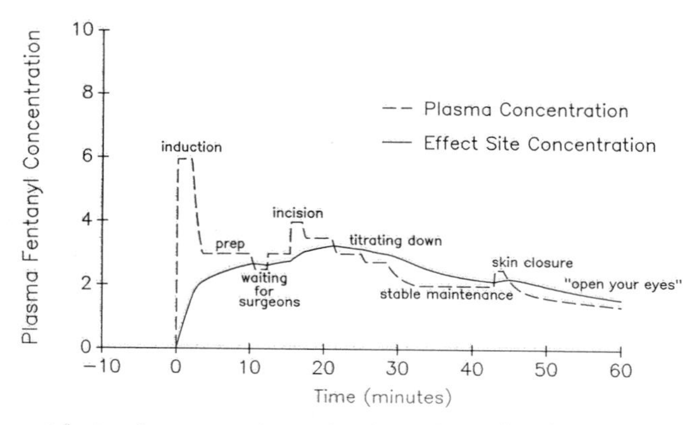

Fig. 4. The apparent fentanyl concentration at the site of drug effect for the anesthetic course shown in Fig. 2.

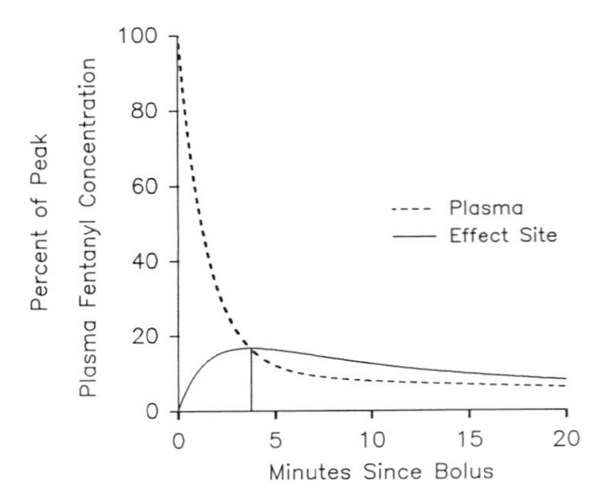

Fig. 5. The fentanyl concentration in the plasma and the apparent fentanyl concentration at the site of drug effect following a bolus injection. The time to the peak concentration at the site of drug effect is marked with a vertical line.

result of the linear nature of the system. Figure 6 shows that proportional increases in bolus size produce proportional increases in the concentration in the plasma and in the apparent concentration at the site of drug effect. However, the overall shape of the curve, including the time of peak concentration, does not change.

A CCIP which targets the apparent concentration at the site of drug effect must account for this hysteresis between the plasma concentrations and the apparent concentrations at the site of drug effect. My approach is to administer the drug so that the apparent concentration of drug at the site of drug effect exactly reaches the desired concentration *as rapidly as possible, without overshooting the desired concentration*. This can only be accomplished by administering a single bolus of drug at time 0, and then waiting for the apparent concentration at the site of drug effect to *coast* up to the desired concentration, as shown in Fig. 7.

It is possible to reach the desired concentration at the site of drug effect more rapidly than the time to peak concentration by administering a larger bolus. As shown in Fig. 8, *overpressuring* the system with a larger bolus to reach the desired concentration at the site of drug effect more rapidly will necessarily produce an

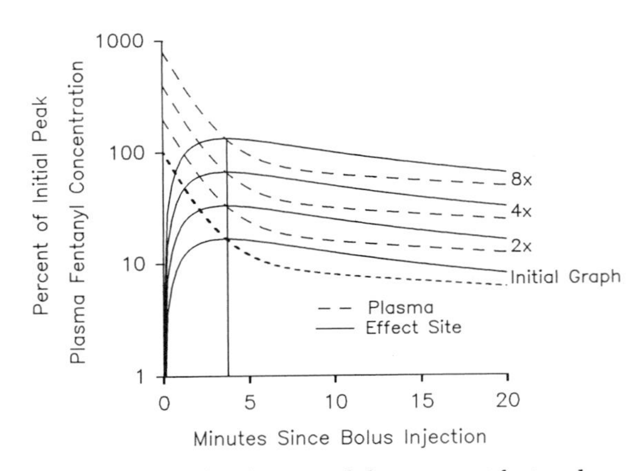

Fig. 6. The fentanyl concentration in the plasma and the apparent fentanyl concentration at the site of drug effect following bolus injections of different doses. The magnitude of the dose does not influence the time to the peak concentration at the site of drug effect.

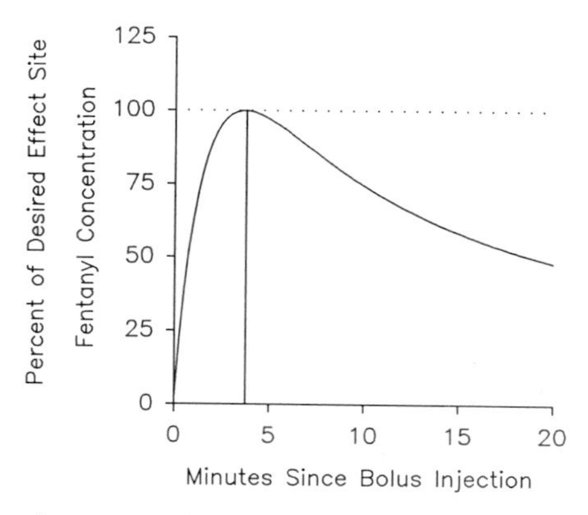

Fig. 7. The apparent fentanyl concentration over time following administration of a bolus injection which exactly reaches the desired target concentration (dotted line) without an overshoot. The time of peak concentration at the site of drug effect is marked with a vertical line.

overshoot of the desired concentration. For the intravenous anesthetics, achieving the desired apparent concentration at the site of drug effect more rapidly than the time to the peak concentration is usually not a sufficiently worthwhile goal to justify overshooting the desired concentration, and may produce significant side effects.

There are infinitely many ways of reaching the desired concentration at the site of drug effect more slowly than the time to the peak concentration from a single bolus. For example, Fig. 9 shows the results of four separate boluses, given two minutes apart, which produce the desired peak effect at 7.75 minutes. Of course, another option is to just give the bolus shown in Fig. 7 four minutes later, also shown in Fig. 9! It is not clear why one would wish to reach the desired concentration more slowly than possible from a single bolus.

Occasionally, the rate of rise of a drug, rather than the absolute concentration, may be a source of toxicity. For example, opioid induced truncal rigidity, which can be life threatening, appears to be caused by the rate of rise in concentration rather than the absolute concentration. However, *imposition of specific limits on rate of rise*

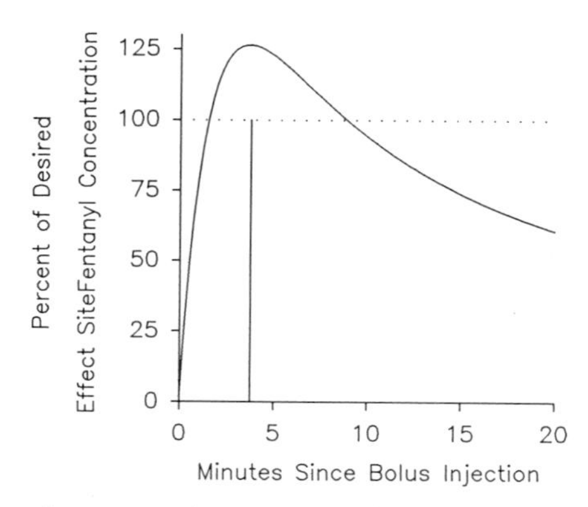

Fig. 8. The apparent fentanyl concentration over time following administration of a bolus injection which reaches the desired target concentration (dotted line) at an earlier time than shown in Fig. 7. After reaching the desired target concentration, the apparent fentanyl concentration at the site of drug effect continues increasing, producing an overshoot. The time of peak concentration at the site of drug effect is marked with a vertical line.

of the apparent concentration at the site of drug effect is a *separate, subsequent* (and very simple) step in the algorithm by which a CCIP computes the final infusion rate. Limits on the infusion rate are imposed only after the CCIP computes the bolus dose necessary to reach the target concentration at the site of drug effect.

Figures 7, 8, and 9 suggest a simple method of reaching a desired concentration at the site of drug effect with a CCIP: administer the bolus whose peak apparent concentration at the site of drug effect is exactly the desired concentration.

Unfortunately, the time to peak effect is only independent of magnitude of the dose when no drug is in the body. Once drug has been administered, the time to peak effect is a function of dose. Figure 10 shows the apparent concentrations at the site of drug effect following two separate boluses. The first bolus is given at time 0, and the second bolus is given at 5 minutes. The apparent concentrations at the site of drug effect following second boluses of differing magnitudes (including 0) are shown by the dashed lines. The apparent concentrations at the site of drug effect from the differing magnitudes of the second bolus, considered without regard to the concentrations resulting from the first bolus, peak at the same time (inset). However, as the apparent concentrations at the site of drug effect following the second bolus are superimposed on the declining concentrations from the first bolus, the resulting peak concentrations do not occur at the same time for the different doses. The horizontal dotted line connects the times of peak concentration at the site of drug effect for second boluses of differing magnitudes. Although the time to peak concentration is less for the second bolus than for the first bolus, it asymptotically approaches the time to peak concentration for the first bolus as the magnitude of the second bolus increases.

Thus, finding the correct bolus to increase the concentration at the site of drug effect to the desired target without overshoot requires solving for both the *magnitude of* and the *time to* the peak apparent drug concentration at the site of drug effect. A numerical search and a closed analytical solution for finding the correct bolus amount have been developed for the drugs with triexponential pharmacokinetics. These will be presented in a subsequent manuscript.

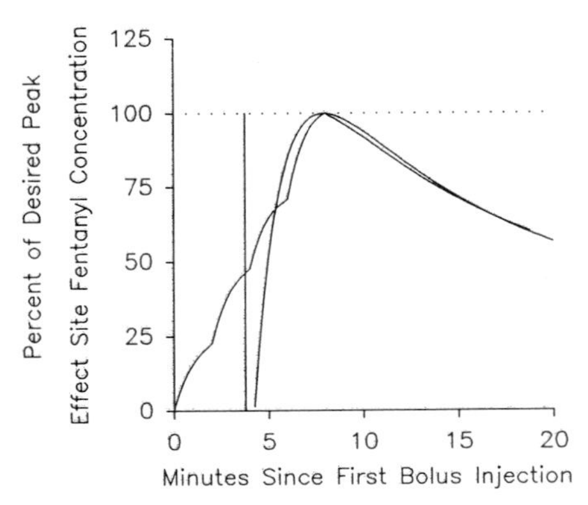

Fig. 9. Two strategies for reaching the desired target concentration (dotted line) at a later time than the time of peak concentration following a bolus injection (indicated by the vertical line). The two curves show the apparent fentanyl concentration at the site of drug effect following four small boluses, and following a single bolus simply given at a later time.

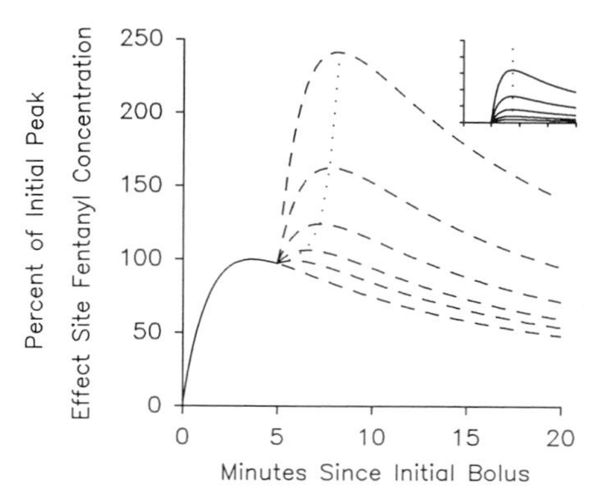

Fig. 10. The apparent fentanyl concentration at the site of drug effect following two boluses, one given at time 0 and one given at 5 minutes. The apparent fentanyl concentrations at the site of drug effect following different doses for the second bolus are shown. The times of peak concentration at the site of drug effect are shown by the dotted line. Although the different doses of the second bolus would have peaked at the same time if no drug were in the body (inset), the superposition of the declining concentrations from the first bolus with the concentrations from the second bolus cause the time to peak concentration at the site of drug effect to be dependent on the magnitude of the second bolus.

After the bolus is administered, no additional drug is given until the concentration at the site of drug effect has risen to the target concentration. At the moment the concentration has peaked at the desired concentration, the apparent concentration at the site of drug effect and the concentration in the plasma are identical. If the CCIP then maintains a constant plasma concentration (a fairly simple task [12,13]), then the apparent concentration at the site of drug effect will also remain constant. To decrease the concentration to a lower target as rapidly as possible, the CCIP must stop the infusion. The CCIP then tracks the decrease in apparent concentration at the site of drug effect until it has decreased to the desired target concentration. At that time, the CCIP can maintain the desired apparent concentration at the site of drug effect by maintaining the same desired concentration in the plasma [12, 13].

Figure 11 shows the plasma concentration and apparent concentration at the site of drug effect for a simple fentanyl infusion regimen targeting the apparent concentration at the site of drug effect. As suggested above, when attempting to increase the apparent concentration at the site of drug effect, the CCIP administers a bolus sufficient to reach the desired concentration, then administers no additional drug until the apparent concentration at the site of drug effect reaches the desired concentration. The CCIP then maintains the concentration in the plasma at the desired concentration, which maintains the apparent concentration at the site of drug effect at the desired target concentration as well. When a second increase is desired, the CCIP again administers a bolus, then waits until the apparent concentration at the site of drug effect reaches the desired target. As initially shown in Fig. 10, this occurs slightly sooner, relative to the time of the second bolus, than following the initial bolus because of the drug already present in the body. Once the apparent concentration at the site of drug effect has peaked at the desired concentration, the CCIP again maintains the apparent concentration at the site of drug effect at the desired concentration by maintaining the plasma concentration at the desired

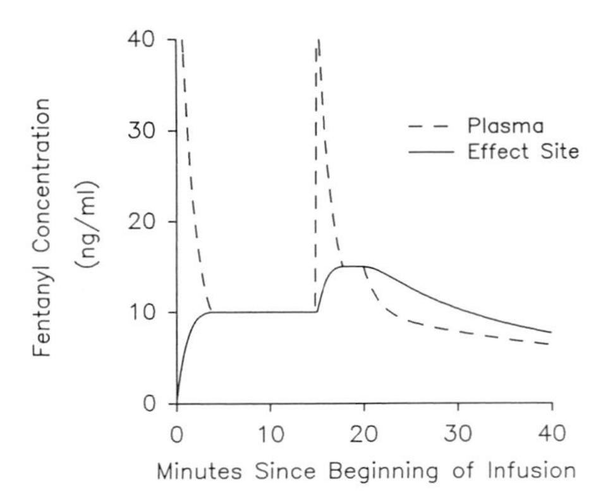

Fig. 11. Fentanyl concentration in the plasma and apparent fentanyl concentration at the site of drug effect for an anesthetic which initially targets a fentanyl concentration (at the site of drug effect) of 10 ng/ml for 15 minutes, then increases the target concentration to 15 ng/ml for 5 minutes, then decreases the target concentration to 8 ng/ml.

concentration. Finally, the CCIP stops the infusion while apparent concentration at the site of drug effect decreases to a lower target concentration. When the apparent concentration at the site of drug effect has decreased to the desired concentration, the CCIP administers a small bolus to increase the plasma concentration to the desired concentration. The CCIP then maintains a steady concentration at the site of drug effect by maintaining a steady plasma concentration.

Figure 12 shows the simulated result obtained when following the same anesthetic course shown in Fig. 2, except that the CCIP targets the apparent concentration at the site of drug effect site rather than the plasma concentration. The rapid response and elegant titration of the plasma drug concentration seen in Fig. 2 cannot be reproduced when targeting the apparent concentration at the site of drug effect. However, this is a reflection of the underlying physiology relating the plasma concentrations to the concentration at the site of drug effect.

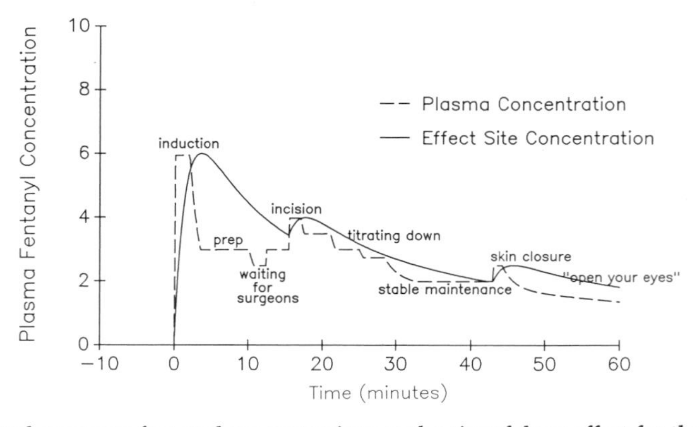

Fig. 12. Simulated apparent fentanyl concentrations at the site of drug effect for the same anesthetic shown in Fig. 2, but with the CCIP targeting the concentration at the site of drug effect.

DISCUSSION

The plasma is not the site of drug effect for the drugs used to induce and maintain the anesthetic state. A CCIP can theoretically provide very precise titration of plasma drug concentration, as shown in Fig. 2. However, this may not translate into precise titration of drug effect, as shown in Fig. 4, because of the time required for transfer of drug from the plasma to the site of drug effect. Thus, titration of the plasma drug concentration may delude the clinician into believing that he or she has more precise control over drug effect than is actually possible.

Since bolus injections into the plasma can increase the plasma concentration to any desired concentration almost instantaneously, a clinician targeting the plasma may believe that the patient has reached a degree of drug effect which has not yet occurred. For example, a clinician using fentanyl prior to intubation might intubate a patient within 1 minute of starting the infusion, as suggested by Fig. 2. Figure 4 clearly shows that at 1 minute the apparent fentanyl concentration at the site of drug effect would be far below the target concentration. Thus, it is probable that the patient would be very lightly anesthetized, or even awake, at the time of intubation! If the apparent concentration at the site of drug effect had been targeted (Fig. 12), and the physician had waited until the CCIP indicated that the target concentration had been reached at the site of drug effect, then the patient would have been adequately anesthetized for the intubation.

Anesthesiologists have routinely used bolus injections to induce anesthesia and to deepen the anesthetic state when their patients appeared lightly anesthetized. When targeting plasma concentrations, a CCIP will administer less drug than anesthesiologists have traditionally used. My personal experience with fentanyl administration by CCIP is that the *quality* of the anesthetic induction, when targeting the plasma concentration, is worse with the CCIP than with the traditional bolus injection. When targeting the plasma concentration, the CCIP starts with an almost trivial fentanyl bolus, then administers an infusion which only produces the desired anesthetic state after 8 to 10 minutes. By contrast, when targeting the apparent concentration at the site of drug effect, the CCIP administers a bolus similar to the traditional induction dose, with an improvement in the quality of the induction. Thus, it would appear that traditional anesthetic practice has been to target the site of drug effect, albeit unknowingly so!

Another major difference between targeting the drug concentration in the plasma and targeting drug concentration at the site of drug effect can be seen when the target concentration is changed to a lower concentration. The rapid distributional pharmacokinetics permit very rapid decreases in plasma concentration for most of the intravenous anesthetics when the infusion is terminated, as shown in Fig. 2. The decrease in apparent concentration at the site of drug effect is much slower, as can be seen in Fig. 12. To decrease the concentration in the plasma to a new target concentration the CCIP will turn off the infusion until the plasma concentration has decreased to the new, lower, target concentration. At this point the CCIP will again infuse drug to maintain the target concentration. This is not logical, because the apparent concentration at the site of drug effect will remain above the target concentration, and will decrease very slowly to the target concentration because the CCIP is again infusing drug. In other words, if the plasma is not the site of drug effect, then it makes no sense to infuse drug into the patient if the apparent concentration at the site of drug effect is higher than the desired concentration. Lastly, an anesthesiologist observing the plasma concentration predicted by a CCIP may not realize that the patient who fails to awaken from the anesthetic when the plasma concentration has decreased appropriate concentrations may be asleep because the apparent concentration at the site of drug effect is substantially higher.

It is interesting to observe that several groups have recently demonstrated a *lack* of correlation between plasma fentanyl concentrations and drug effect, when infusing fentanyl using a CCIP that targets the concentration in the plasma [14,15]. The disequilibrium between the plasma and the site of drug effect may explain the inability of these studies to demonstrate a concentration-effect relationship.

In summary, this manuscript has outlined an algorithm for a CCIP to increase, maintain, and decrease the apparent drug concentration at the site of drug effect. The detailed mathematics of the algorithm will be presented in a subsequent manuscript. Despite the great attention given to targeting plasma drug concentrations with CCIPs, some of this effort may have been fundamentally misdirected when the plasma is not the site of drug effect.

ACKNOWLEDGMENTS

Supported in part by the Veterans Administration Merit Review Program, a Starter Grant from the American Society of Anesthesiologists, and Biomedical Research Support Grant RR05353 awarded by the Biomedical Research Support Grant Program, Division of Research Resources, National Institutes of Health, and the Anesthesia/Pharmacology Research Foundation.

REFERENCES

1. E. Krüger-Thiemer. Continuous intravenous infusion and multicompartment accumulation. *Eur. J. Clin. Pharmacol.* 4:317-324 (1968).
2. H. Schwilden. A general method for calculating the dosage scheme in linear pharmacokinetics. *Eur. J. Clin. Pharmacol.* **20**:379-383 (1981).
3. J. M. Alvis, J. G. Reves, A. V. Govier, P. G. Menkhaus, C. E. Henling, J. A. Spain, and E. Bradley. Computer-assisted continuous infusions of fentanyl during cardiac anesthesia: comparison with a manual method. *Anesthesiology* **63**:41-49 (1985).
4. J. M. Alvis, J. G. Reves, J. A. Spain, and L. C. Sheppard. Computer-assisted continuous infusion of the intravenous analgesic fentanyl during general anesthesia-an interactive system. *IEEE T. Bio-med. Eng.* **32**:323-329 (1985).
5. M. E. Ausems, D. R. Stanski, and C. C. Hug. An evaluation of the accuracy of pharmacokinetic data for the computer-assisted infusion of alfentanil. *Brit. J. Anaesth.* **57**:1217-1225 (1985).
6. S. L. Shafer, J. R. Varvel, N. Aziz, and J. C. Scott. The pharmacokinetics of fentanyl administered by computer-controlled infusion pump. *Anesthesiology* (in press).
7. D. B. Raemer, A. Buschman, J. R. Varvel, B. K. Philip, M. D. Johnson, D. A. Stein, and S. L. Shafer. The prospective use of population pharmacokinetics in a computer-driven infusion system for alfentanil. *Anesthesiology* **73**:66-72 (1990).
8. J. C. Scott, D. R. Stanski, S. Vozeh S. ,R. D. Miller, and J. Ham. Simultaneous modeling of pharmacokinetics and pharmacodynamics: Application to d-tubocurarine. *Clin. Pharmacol. Ther.* **25**:358-371 (1979).
9. L. B. Sheiner, D. R. Stanski, S. Vozeh S. ,R. D. Miller, and J. Ham. Simultaneous modeling of pharmacokinetics and pharmacodynamics: Application to d-tubocurarine. *Clin. Pharmacol. Ther.* **25**:358-371 (1979).
10. C. J. Hull, H. B. Van Beem, K. McLeod, A. Sibbald, and M. J. Watson. A pharmacodynamic model for pancuronium. *Brit. J. Anaesth.* **50**:1113-1123 (1978).
11. J. C. Scott, K. V. Ponganis, and D. R. Stanski. EEG quantitation of narcotic effect: the comparative pharmacodynamics of fentanyl and alfentanil. *Anesthesiology* **62**:234-241 (1985).

12. S. L. Shafer, L. C. Siegel, J. E. Cooke, and J. C. Scott. Testing computer-controlled infusion pumps by simulation. *Anesthesiology* **68**:261-266 (1988).
13. P. O. Maitre and S. L. Shafer. A simple pocket calculator approach to predict anesthetic drug concentrations from pharmacokinetic data. *Anesthesiology* **73**:332-336 (1990).
14. P. S. A. Glass, J. R. Jacobs, L. R. Smith, B. Ginsberg, S. A. Bai, and J. G. Reves. Pharmacokinetic model-driven infusion of fentanyl. *Anesthesiology* (in press).
15. I. Schewiger, J. Bailey, and C. C. Hug Jr. Anesthetic interactions of computer-controlled and constant rate infusions of fentanyl and midazolam for cardiac surgery. *Anesthesiology* **71**:A305 (1989).

CONTRIBUTORS

Janet Anderson, Department of Pharmacy, Western Infirmary, Glasgow, Scotland G11 6NT, United Kingdom.

Gordon L. Amidon, College of Pharmacy, The University of Michigan, Ann Arbor, Michigan 48109–1065.

Darrell R. Abernethy*, Program in Clinical Pharmacology, Brown University, Roger Williams General Hospital, 825 Chalkstone Avenue, Providence, Rhode Island 02908.

J. Douglas Briggs, Department of Renal Medicine, Western Infirmary, Glasgow, Scotland G11 6NT, United Kingdom.

Lloyd W. Burgess, Center for Process Analytical Chemistry, BG-10, University of Washington, Seattle, Washington 98195.

Haiyung Cheng*, Merck Sharp & Dohme Research Laboratories, Sumneytown Pike, West Point, Pennsylvania 19486; and Department of Pharmaceutics, School of Pharmacy, State University of New York, Buffalo, New York 14260.

Win L. Chiou*, Department of Pharmacodynamics, College of Pharmacy, The University of Illinois at Chicago, (M/C 865), Chicago, Illinois 60612.

David Z. D'Argenio*, Department of Biomedical Engineering, University of Southern California, Los Angeles, California 90089–1451.

William E. Evans, Pharmacokinetics and Pharmacodynamics Section, Pharmaceutical Division, St. Jude Children's Research Hospital, 332 N. Lauderdale, Memphis, Tennessee 38101–0318; and The Center for Pediatric Pharmacokinetics and Therapeutics, Departments of Clinical Pharmacy and Pediatrics, University of Tennessee, Memphis, Tennessee 38163.

James M. Gallo*, Department of Pharmaceutics, College of Pharmacy, University of Georgia, Athens, Georgia 30602.

Nicholas H.G. Holford*, Department of Pharmacology, University of Auckland, Auckland, New Zealand.

William J. Jusko*, Department of Pharmaceutics, School of Pharmacy, State University of New York, Buffalo, New York 14260.

Andrew W. Kelman, Department of Medicine and Therapeutics, University of Glasgow, Stobhill General Hospital and Western Infirmary, Glasgow, Scotland G21 3UW, United Kingdom.

Alain Mallet, Méthodologie Informatique et Statistique en Médecine, SIM-INSERM U 194, 91, bd de l'Hôpital, 75634 Paris Cedex 13, France.

France Mentré*, Méthodologie Informatique et Statistique en Médecine, SIM-INSERM U 194, 91, bd de l'Hôpital, 75634 Paris Cedex 13, France.

Sabina Merlo, Center for Bioengineering, FL–20, University of Washington, Seattle, Washington 98195; and The Washington Technology Center, Seattle, Washington 98195. Current address: Dipartimento di Elettronica, Universita' di Pavia, Via Abbiategrasso 209, 27100 Pavia, Italy.

Angela Munday, Department of Pharmacy, Western Infirmary, Glasgow, Scotland G11 6NT, United Kingdom.

Alison A. Niven, Department of Medicine and Therapeutics, University of Glasgow, Stobhill General Hospital and Western Infirmary, Glasgow, Scotland G21 3UW, United Kingdom.

Doo-Man Oh, College of Pharmacy, The University of Michigan, Ann Arbor, Michigan 48109–1065.

John H. Rodman*, Pharmacokinetics and Pharmacodynamics Section, Pharmaceutical Division, St. Jude Children's Research Hospital, 332 N. Lauderdale, Memphis, Tennessee 38101–0318; and The Center for Pediatric Pharmacokinetics and Therapeutics, Departments of Clinical Pharmacy and Pediatrics, University of Tennessee, Memphis, Tennessee 38163.

Steven L. Shafer*, Department of Anesthesia, Stanford University School of Medicine, Stanford, California 94305; and Veterans Administration Medical Center, 3801 Miranda Avenue, Palo Alto, California 94304.

Alan Schumitzky*, Department of Mathematics, University of Southern California, Los Angeles, California 90089–1113.

Patrick J. Sinko*, College of Pharmacy, The University of Michigan, Ann Arbor, Michigan 48109–1065; and Therapeutics Systems Research Laboratories, Inc., P.O. Box 7062, Ann Arbor, Michigan 48107.

Donald R. Stanski*, Department of Anesthesia, Stanford University School of Medicine, Stanford, California 94305; and Veterans Administration Medical Center, 3801 Miranda Avenue, Palo Alto, California 94304.

George D. Swanson*, Anesthesiology Department, University of Colorado Medical School, Denver, Colorado 80262; and Department of Physical Education, California State University, Chico, California 95929.

Alison H. Thomson, Department of Medicine and Therapeutics, University of Glasgow, Stobhill General Hospital and Western Infirmary, Glasgow, Scotland G21 3UW, United Kingdom.

Michael Weiss*, Institut für Pharmakologie und Toxikologie, Martin-Luther-Universität Halle-Wittenberg, Postfach 302, 4020 Halle, Germany.

Brian Whiting*, Department of Medicine and Therapeutics, University of Glasgow, Stobhill General Hospital and Western Infirmary, Glasgow, Scotland G21 3UW, United Kingdom.

Paul Yager*, Center for Bioengineering, FL–20, University of Washington, Seattle, Washington 98195; and The Washington Technology Center, Seattle, Washington 98195.

Ruomei Zhang, Department of Biomedical Engineering, University of Southern California, Los Angeles, California 90089–1451.

* Author to whom correspondence should be addressed.

PARTICIPANTS

Peter H. Abbrecht
Uniformed Services University
of the Health Sciences
Bethesda, MD 20814–4799

Darrell R. Abernethy
Brown University
Providence, RI 02908

Andrew A. Acheampong
Allergan Inc.
Irvine, CA 92715

Gordon L. Amidon
The University of Michigan
Ann Arbor, MI 48109-1065

Mordechai Ben Porat
Israel Governmental Research Labs.
Petach – Tikva 49558
Israel

Sven Björkman
V.A. Medical Center
Palo Alto, CA 94304

Clive E. Bowman
PROCESS
Berks, SL6 5DE
United Kingdom

Ricardo Brechner
University of Southern California
Los Angeles, CA 90033

Kenneth Brouwer
University of North Carolina
Chapel Hill, NC 27599

Kim L.R. Brouwer
University of North Carolina
Chapel Hill, NC 27599

Ann Chang
University of California
Los Angeles, CA 90024

Benjamin C. Chen
University of California
Los Angeles, CA 90024

Haiyung Cheng
Merrell Dow Pharmaceuticals, Inc.
Indianapolis, IN 46268

Mary H. Cheng
Children's Hospital of Los Angeles
Los Angeles, CA 90027

Du–Shieng Chien
Allergan Pharmaceuticals
Irvine, CA 92715

Win L. Chiou
University of Illinois at Chicago
Chicago, IL 60680

Arthur K. Cho
University of California
Los Angeles, CA 90024

Teresa Chu
Los Angeles, CA 90025

Thomas G. Coleman
University of Mississippi Med. Center
Jackson, MS 39216–4505

Kenneth Courtney
Palo Alto Medical Foundation
Palo Alto, CA 94301

Pamela G. Coxson
Lawrence Berkeley Lab.
Berkeley, CA 94720

Fran Cunningham
University of Illinois
Chicago, IL 60612

David Z. D'Argenio
University of Southern California
Los Angeles, CA 90089–1451

Carlo De Angeles
Sunnybrook Health Science Centre
North York, Ontario
Canada M4N 3M5

Charles Denaro
San Francisco General Hospital
San Francisco, CA 94110

Barry Dick
Palo Alto Veterans Medical Center
Palo Alto, CA 94304

Joseph J. DiStefano, III
University of California, L.A.
Los Angeles, CA 90024–1596

Michael Dudley
Brown University
Providence, RI 02908

William F. Ebling
Palo Alto Veterans Medical Center
Palo Alto, CA 94304

James H. Fischer
Lemont, IL 60439

John S. Fitzpatrick
University of Southern California
Los Angeles, CA 90089–1451

Carolyn M. Fleming
Vanderbilt University
Nashville, TN 37122

Roberto Foroni
USLL 25
37122 Verona
Italy

James M. Gallo
University of Georgia
Athens, GA 30602

Douglas Geraets
The University of Iowa
Iowa City, IA 52242

Thomas Gilman
University of Southern California
Los Angeles, CA 90033

Pilar Gomis
University of Southern California
Los Angeles, CA 90033

Mark Gumbleton
University of California, San Francisco
San Francisco, CA 94143

Victoria Hale
Baltimore, MD 21210

Charles E. Halstenson
Hennepin County Medical Center
Minneapolis, MN 55415

Li–Min Haung
University of California
Los Angeles, CA 90024–1596

Thomas K. Henthorn
Northwestern University
Chicago, IL 60611

Kenneth J. Himmelstein
Allergan
Irvine, CA 92715

Nicholas H.G. Holford
University of Auckland
Auckland
New Zealand

Phillip Hong
University of Southern California
Los Angeles, CA 90033

Agneta Hurst
University of Southern California
Los Angeles, CA 90033

George Jaresko
University of Southern California
Los Angeles, CA 90033

Roger W. Jelliffe
University of Southern California
Los Angeles, CA 90033

Donald T. Jung
SYNTEX
Palo Alto, CA 94304

Robert E. Kates
Analytical Solutions, Inc.
Sunnyvale, CA 94089

Darryl Katz
California State University
Fullerton, CA 92634

Toshihiro Kikkoji
University of California
San Francisco, CA 94143

Thomas H. Kramer
College of Medicine
Tucson, AZ 85724

Yoshito Kumagai
University of Southern California
Los Angeles, CA 90033

Masahiro Kurita
University of California
La Jolla, CA 92093–0813

Elliot M. Landaw
University of California
Los Angeles, CA 90024–1766

David Lau
University of California
San Francisco, CA 94143–0446

Frank Lee
Glaxo Incorporated
RTP, NC 27709

Vahn A. Lewis
University of Texas
Houston, TX 77225

Lena Y. Lin
University of California
Los Angeles, CA 90024

Min Lin
University of Southern California
Los Angeles, CA 90033

Shu–Ching Lin
University of Illinois
Chicago, IL 60612

Thomas Ludden
University of Texas Health Sci. Center
San Antonio, TX 78284

Daniel C. Maneval
Genentech, Inc.
South San Francisco, CA 94080

Etan Markowitz
California State University
Fullerton, CA 92634

Richard M. Matsumoto
Allergan Pharmaceuticals
Irvine, CA 92715

France Mentre
INSERM U194
F–75643 Paris Cedex 13
France

Gary Milavetz
The University of Iowa
Iowa City, IA 52242

Malcolm J. Moore
Princess Margaret Hospital
Toronto, Ontario
Canada M4X 1K9

Maulik Nanavaty
Baxter Healthcare Corporation
Deerfield, IL 60015

Chin K. Ng
University of California
Los Angeles, CA 90024

Ann Nguyen
University of California
San Francisco, CA 94143

Thuvan T. Nguyen
University of California
Los Angeles, CA 90024–1596

Shigehiro Ohdo
Alhambra, CA 91801

Joseph Ojingwa
University of California
San Francisco, CA 94143–0446

Sasa Pajevic
University of California
Los Angeles, CA 90024

John G. Pierce
California State University
Fullerton, CA 92634

Beatrice Perotti
University of California
San Francisco, CA 94143

Gary M. Pollack
University of North Carolina
Chapel Hill, NC 27599

J.H. Proost
University of Groningen
9713 AW Groningen
The Netherlands

Thomayant Prueksaritanont
University of California
San Francisco, CA 94143–0446

David Raffel
University of Wisconsin
Madison, WI 53706

Srinivas Ramamurthy
University of California
Los Angeles, CA 90024

Mary Relling
St. Jude Children's Research Hospital
Memphis, TN 38101

Joseph Richman
Allergan Pharmaceuticals
Irvine, CA 92715

Vincent C. Rideout
University of Wisconsin
Madison, WI 53706

John H. Rodman
St. Jude Children's Research Hospital
Memphis, TN 38105

Brian M. Sadler
Research Triangle Institute
Research Triangle Park, NC 27709

Ronald J. Sawchuk
University of Minnesota
Minneapolis, MN 55455

Olga Schmidlin
University of California
San Francisco, CA 94143–0446

Debra Schmitz
Children's Hospital of Los Angeles
Los Angeles, CA 90054

Roy W. Schubert
Louisiana Tech University
Ruston, LA 71272

Alan Schumitzky
University of Southern California
Los Angeles, CA 90089-1113

Helmut Schwilden
Universität Bonn
D-5300 Bonn 1,
Germany

Steven L. Shafer
Veterans Administration Med. Center
Palo Alto, CA 94304

Jaymin Shah
Syntex A3–165
Palo Alto, CA 94303

Lewis B. Sheiner
University of California
San Francisco, CA 94143

Ru–Liang Shih
University of Southern California
Los Angeles, CA 90033

Ronald A. Siegel
University of California
San Francisco, CA 94143–0446

John B. Slate
MiniMed Technologies
Sylmar, CA 91342

Gisela Spieler
California State University
Fullerton, CA 92634–9480

S.P. Srinivas
Stanford University
Stanford, CA 94305

Donald R. Stanski
Veterans Administration Med. Center
Palo Alto, CA 94304

J.H. Steinbach
Veterans Administration Med. Center
San Diego, CA 92161

John St. Peter
Hennepin County Medical Center,
Minneapolis, MN 55415

Helen Sun
University of California
San Francisco, CA 94143–0446

George D. Swanson
California State University, Chico
Chico, CA 95929

Diane Tang–Liu
Allergan, Inc.
Irvine, CA 92715

Mary Teresi
The University of Iowa
Iowa City, IA 52242

Jake J. Thiessen
University of Toronto
Toronto, M5S 1A1
Canada

Hugo Vargas
University of California
Los Angeles, CA 90024–1735

Jan Volhejn
MEDISOFT
100 00 Prague
Czechoslovakia

D. Russell Wada
University of California
Los Angeles, CA 90024

Parker Waechter
University of California
Los Angeles, CA 90024–1596

Jonathan Wakefield
Nottingham University
Nottingham NG7 2RD
United Kingdom

Jeffrey Wald
State University of New York
Buffalo, NY 14260

Michael Weiss
Martin Luther Universität
4020 Halle
Germany

Brian Whiting
University of Glasgow
Glasgow, G21 3UW
United Kingdom

Paul Yager
University of Washington
Seattle, WA 98195

Wan–Ching Yen
University of Southern California
Los Angeles, CA 90033

Dan–Chu Yu
University of California
Los Angeles, CA 90024

Roumei Zhang
University of Southern California
Los Angeles, CA 90089–1451

INDEX